Environmental Remediation Technologies, Regulations and Safety

nova
science publishers

www.novapublishers.com

Environmental Remediation Technologies, Regulations and Safety

Wind Turbines: Technology, Applications and Efficiency
David Borge-Diez, PhD
and Daniel Orlando Icaza-Álvarez (Editors)
2022. ISBN: 978-1-68507-974-1 (Hardcover)
2022. ISBN: 979-8-88697-159-0 (eBook)

Phthalates: Environmental and Health Effects
Müfide Aydoğan Ahbab, PhD
and Nurhayat Barlas, PhD (Editors)
2022. ISBN: 978-1-68507-970-3 (Hardcover)
2022. ISBN: 979-8-88697-038-8 (eBook)

Radium in the Hydrosphere of Brazilian Alkaline Areas
Luis Henrique Mancini, PhD and Daniel Marcos Bonotto (Editors)
2022. ISBN: 978-1-68507-889-8 (Hardcover)
2022. ISBN: 979-8-88697-009-8 (eBook)

Organophosphates: Detection, Exposure and Occurrence. Volume 2: Acute Exposure and Treatments
Tamara Lazarević-Pašti, PhD (Editor)
2022. ISBN: 978-1-68507-659-7 (Hardcover)
2022. ISBN: 978-1-68507-734-1 (eBook)

Investigations into the Decarbonisation of Railways
Inara Watson, PhD
2021. ISBN: 978-1-68507-250-6 (eBook)

More information about this series can be found at
https://novapublishers.com/product-category/series/environmental-remediation-technologies-regulations-and-safety/

Neha Agarwal
Editor

Pharmaceuticals

Boon or Bane

nova
science publishers

www.novapublishers.com

DOI: https://doi.org/10.52305/GPMC4427

NOTICE TO THE READER

The Publisher has taken reasonable care in the preparation of this book, but makes no expressed or implied warranty of any kind and assumes no responsibility for any errors or omissions. No liability is assumed for incidental or consequential damages in connection with or arising out of information contained in this book. The Publisher shall not be liable for any special, consequential, or exemplary damages resulting, in whole or in part, from the readers' use of, or reliance upon, this material. Any parts of this book based on government reports are so indicated and copyright is claimed for those parts to the extent applicable to compilations of such works.

Independent verification should be sought for any data, advice or recommendations contained in this book. In addition, no responsibility is assumed by the Publisher for any injury and/or damage to persons or property arising from any methods, products, instructions, ideas or otherwise contained in this publication.

This publication is designed to provide accurate and authoritative information with regard to the subject matter covered herein. It is sold with the clear understanding that the Publisher is not engaged in rendering legal or any other professional services. If legal or any other expert assistance is required, the services of a competent person should be sought. FROM A DECLARATION OF PARTICIPANTS JOINTLY ADOPTED BY A COMMITTEE OF THE AMERICAN BAR ASSOCIATION AND A COMMITTEE OF PUBLISHERS.

Additional color graphics may be available in the e-book version of this book.

Library of Congress Cataloging-in-Publication Data

ISBN: 979-8-88697-487-4

Published by Nova Science Publishers, Inc. † New York

I am happy to recommend the book "Pharmaceuticals: Boon or Bane" for all because it is based on the current environmental issues related to the release and accumulations of pharmaceuticals and their toxic effects on human health, environment and ecology. The entire contents of the book are highly useful not only for the researchers, academicians and students but also for general public. The book also covers harmful post effects of extensive use of antibiotics during COVID-19 pandemic and proposes the possible ways of removal of these pharmaceutical contaminants from the environment.

Dr. Namrata Rastogi
Principal Scientist
CSIR-Cental Drug Research Institute
Lucknow, India

Pharmaceuticals: Boon or Bane is very informative, understandable, and useful guide full of important information for researchers, academicians, and students to understand the repercussions behind the extensive production and usage of pharmaceuticals. I am sure this book is going to be of great interest among scientific community and to create awareness for prescribed and rational usage of drugs for a healthier life and safe environment.

Brijesh Pare
President -Association of Chemistry Teachers, India

It's my great pleasure to write an endorsement for the book "Pharmaceuticals: Boon or Bane" which addresses most of the issues, risks, and concerns related to the release of pharmaceuticals into the environment. This is an interesting book for the readers who want to learn about the role of pharma-related contaminants, their exposures, and possible solutions to degrade these contaminants efficiently to save our environment and health. As a pediatric cancer biologist, I highly recommend this book to readers to enhance their knowledge as these contaminants affect our infants and young children.

Ranjana Km Kanchan
CSIR-SRF
Toxicology Division
Cdri, Lucknow, India

I inform you that the book "Pharmaceuticals: Boon or Bane" is written in a very simple and easy language and is suitable according to the needs of the present scenario. This book adequately focuses on environmental issues due to accumulation of pharmaceutical contaminants, human health hazards due to continuous and over usage of pharmaceutical compounds and its side effects. Furthermore, it is also focused on the possible ways to degrade these compounds after their release into the environment and obtain robust and cost efficient techniques. The contents of this book are of high quality and the entire syllabus is based on pharmaceutical compounds. This book will also be very useful for those who are doing research on pharmaceutical compounds. I, Dr. R. R. Awasthi highly recommend the book.

Dr. R. R. Awasthi
Assistant Professor
Department of Physics
Khwaja Moinuddin Chishti University
Lucknow, India

Contents

Foreword

Pharmaceutical analysis is a branch of practical chemistry that involves a series of process for identification, determination, quantification and purification of a substance, separation of the components of a solution or mixture, or determination of structure of chemical compounds. Pharmaceuticals form a diversified group of bioactive ingredient like antiseptics, antipyretics, disinfectants; though they are used for the well-being of everybody, their continuous presence in the environment in the form of metabolic byproducts has caused serious dangers especially wide ranging pollution in the whole environment.

Over the last fifteen to twenty five years there has been progress in several scientific fields, particularly combinatorial Chemistry, genomics, proteomics and bioinformatics which are promises for the future in streamlining of procedures discovering new drugs. The main idea of new methods of research and drug development is to identify the biological action, gene, or protein that disrupted in the disease process. Pollution is a serious problem ever since sewage and industrial effluents are disposed into water courses and on land. It has increased with the growth of industry as well as of population. As a result, it has posed serious threats to human health and eco system. Industries need a wide variety of raw materials and chemicals which are later discharged via their effluents. These effluents contain a wider variety of materials of both organic and inorganic nature including toxic substances and are usually discharged with or without treatment into surface waters such as rivers, streams, lakes or into oceans or on land on into sewers. The solid wastes produced in these industries are not given so much importance as they are recycled, dumped or disposed of to remote places. Liquid industrial wastes are of great concern because of their harmful effects.

Several groups of Indian scientists have studied about the hazardous materials flowing through industrial wastes. Moreover, adverse and side effects of industrial drain effluents on men, flora and fauna have been of immense interest to the scholars of India as well as abroad. In recent years, pharmaceutical studies have gained much popularity.

The present book entitled Pharmaceuticals: Boon or Bane is a brilliant attempt in the field of pharmaceutical studies. The editor of the book, Dr. Neha Agrawal deserves kudos for her sincere efforts including well written papers in the book. She has included twelve chapters related to the varied nuances of the pharmaceuticals. The areas covered in the book range from release and accumulation of pharmaceuticals to their impact on eco-system; from the industrial pollution to emerging trends in pharmaceutical waste generation and from the study of pharmaceutical drugs to the treatment of pharmaceutical water bodies.

Moreover, the editor has left no stone unturned in making the book all comprehensive encompassing all the major aspects related to the study of pharmaceuticals. She has explored each and every minutest detail related to the subject matter with a great deal of critical insight and understanding. Her attempt is quite praiseworthy. I regard the work as scholarly and hope that the readers in general and those who have an abiding interest in pharmaceutical studies in particular would find it extremely rewarding. My good wishes to the editor for her appreciable attempt. She deserves thanks for the work.

Janmashtami
19th August, 2022

Dr. Neeraj Kumar
Professor
P.G. Dept. of English
Magadh University
Bodh-Gaya, Bihar, India

Preface

Pharmaceuticals form a diversified group of bioactive ingredients such as antibiotics, analgesics, antipyretics, disinfectants, antiseptics, antidepressants whose trace presence in global environment has raised concerns on their toxic nature and their post implications on environment and ecosystem. Though used for the well being of everyone their continuous release in the aquatic environment in the form of metabolic by-products and active pharmaceutical ingredients due to irrational usage, commercial activities has imposed chronic effects on human health, marine life and animals.

Pharmaceuticals: Boon or Bane! is a compilation of various chapters on contamination of environment due to release of different categories of drugs, toxic effects of antibiotics, environmental and ecological risk assessment due to existence of drug metabolites, treatment techniques and bioremediation of contaminated waste water to attain a sustainable and green environment. The book will be highly useful for educators, researchers, academicians and for the general public who is unaware of the repercussions of the extensive and irrational usage of drugs.

Agarwal et al. in chapter one has discussed about the pathways of pharmaceutical metabolites and transformation products into the global environment and their toxic effects on human health and environment. They have also highlighted the possible actions and ways which could reduce pharmaceutical environmental contaminants for a healthier future.

Yadav et al. in chapter two have given a comprehensive description of evolutionary, historical and structural aspects of commonly used pharmaceuticals worldwide along with their mechanism of action in human body. They have also thrown light on the clinical usage, and possible side effects of various classes of medicines.

Gupta et al. in chapter three have given an overview of the pollution caused by pharma industries, generation of pharmaceutical waste. They have given detailed information on sources and occurrences of different drugs worldwide with a special focus on the adverse effects caused due to pollution generated by pharma industries.

Yadav et al. in chapter four have given a detailed account of unused antibiotics and their presence in soil and water, types of biomedical waste leading to the antibiotic-resistant issue. They have suggested the drastic change in pharmaceutical waste before and after COVID 19 pandemic, color coding for the storage of biomedical waste along with disposal methods to minimize risks.

Singh Anupma in chapter five has summarized the adverse effects associated with the long term usage of analgesics on human health, their pharmacokinetics and interaction with other drugs, their impacts on human health and environment.

Ghosh et al. in chapter six have studied the Diclofenac economy in terms of its growth and related chemicals manufacturing and marketing in different regions of India, its indirect toxic effects on human beings, vultures and marine life especially fishes. They have addressed the present and future strategies to reduce the toxic effects by enhancing diclofenac elimination from the global market.

Singh et al. in chapter seven have elucidated a case study of cocryatals of paracetamol, its idea and advantages. Through their study they have demonstrated the need for future research in the field of pharmaceutical cocryatals to improve existing medications and its supply.

In chapter eight Gupta et al. identified and summarized different sources of pharmaceutical pollution and its adverse effects on various components of ecology. Mechanism of pharmaceuticals induced antimicrobial resistance and its harmful effects on aquatic environment have been elaborated in this chapter. Lastly, the technologies available for removal of pharmaceutical pollutants and their efficiency have been discussed.

Solanki et al. in chapter nine have given a detailed account of pharmaceuticals used during COVID-19 pandemic and also suggested the green and sustainable approach to remove pharmaceutical contaminants from the aquatic environment by heterogenous photocatalytic mineralization technique in presence of LED irradiations.

Sushmita Bajpai in chapter ten has attempted to provide a holistic viewpoint on different types of waste water effluents generated by various industrial and non-industrial sectors and their treatment techniques in water bodies along with policy instruments to control the discharge of pharmaceutical instruments.

Agrawal et al. in chapter eleven have highlighted the latest trends for the safe removal of pharmaceutical contaminants from the global environment through green and sustainable approach. They have also discussed the limitations of bioremediation of contaminants and future prospects in further research for a green and safe environment.

Singh et al. in chapter twelve have given a broad review on different types of metal oxide nanomaterials and their applications in pharmaceuticals and various other fields of nanotechnology along with gas sensing in the field of pharmaceuticals, batteries, magnetic storage media, catalyst, various types of solar cells, energy conversion devices, rectifiers, electronic and optoelectronics devices have also been discussed.

Acknowledgments

With profound sense of reverence and gratitude, I express my hearty indebtedness to Nova Science Publishers, Inc. for their specialized support and encouragement given to me for completing this project and for publishing it in the form of edited book.

I would like to acknowledge the contribution of all the authors for making this publication successful. Their research and input were essential for the successful completion of this work.

I wish to pay heartfelt gratitude to my teachers and mentors for their contribution to my work and I most gratefully acknowledge the constant encouragement and support given by my friends and family members for their unconditional and whole hearted support to complete this work successfully.

I wish to express my loving feelings to my children who supported me by their constant affection, understanding and moral support which enabled me to complete this work without any hurdle.

Finally, I would like to thank The Great Almighty who has always been there to bless me to achieve success in all my endeavors. This achievement would not have been possible without his mercy.

Thank you GOD, thank you all.

Dr. Neha Agarwal
Editor

Chapter 1

Release and Accumulation of Pharmaceuticals in the Environment: Critical Risk Assessment and Challenges for Environment, Ecosystem and Human Health

Neha Agarwal[1,*], Vijendra Singh Solanki[2], R. R. Awasthi[3], Vimala Bind[4], Bhawana Edison[5], Mamta Verma[1] and Virendra Kumar[6]

[1]Department of Chemistry, Navyug Kanya Mahavidyalaya, Lucknow, India
[2]Department of Chemistry, Mody University of Science and Technology, Rajasthan, India
[3]Department of Physics, University of Lucknow, Lucknow, India
[4]Department of Zoology, Navyug Kanya Mahavidyalaya, Lucknow, India
[5]Department of Botany, Navyug Kanya Mahavidyalaya, Lucknow, India
[6]Department of Chemistry, Feroze Gandhi College, Raibareili, India

Abstract

Pharmaceuticals are essential to ensure high standards of human health and well-being. Extensive and uncontrolled usage under different classes, such as antibiotics, analgesics, antipyretics, antidepressants, anti-convulsants, beta-blockers, steroids, antihypertensive drugs etc., poses high environmental and human health hazards when exposed to and accumulated in the environment and ecosystem. In recent years, the accumulation of pharmaceutical ingredients (as emerging pollutants, intermediates, and raw materials) in the environment has received great attention all over the world due to their frequent detection in aquatic environments as subclasses of organic contaminants. Still, there is a lack

* Corresponding Author's Email: nehaagarwal4074@gmail.com.

In: Pharmaceuticals: Boon or Bane
Editor: Neha Agarwal
ISBN: 979-8-88697-487-4

of knowledge concerning pharmaceutical quantities and their metabolites entering the environment, active pharmaceutical ingredients (APIs), metabolism and transformation pathways, and their persistence or biodegradability in the environment. Human health is directly or indirectly affected by pharmaceutical effluents, especially in the vicinity of pharmaceutical industrial zones due to the greater probability of contamination in areas of proximity, including drinking water. Latest epidemiological studies predict the possibility of relative risks of brain disorders increasing due to paracetamol exposure.

The study of potential impacts of the toxicity of pharmaceutical byproducts on the environment and human health has been prioritized in this chapter. It highlights risks pertaining to health and assesses environmental hazards due to contamination of the global environment with pharmaceutical pollutants. Traditional treatment methods of wastewater are not sufficient for the eradication of active pharmaceutical ingredients and their metabolites from aquatic environments; whereas advanced treatment techniques are not sustainable because of energy consumption, high operational cost, efficiency, and efficacy. Therefore, the appropriateness for extensive pharmaceutical production and its rational usage must be checked at the source, including a shift to the production of better biodegradable pharmaceuticals in the long run for "sustainable pharmacy."

Keywords: pharmaceutical effluents, active pharmaceutical ingredients, health hazards, risk assessment, sustainable pharmacy

1. Introduction

On a global level, pharmaceuticals are emerging as one of the most hazardous global apprehensive pollutants due to their frequent detection, especially in aquatic environments and ecosystems [1-4]. The first evidence of their existence was detected in the 1970s [5]. Since then, pharmaceuticals have been detected in different samples taken from aquatic environments such as ponds, river streams, groundwater aquifers, drinking water, sewage-wastewater treatment plants (WWTPs), and surface waters [6-8]. These compounds are being used for treatment purposes, and simultaneously, the pharmaceutical industries have been releasing highly toxic contaminants into the environment directly or indirectly after chemical modifications. Pharmaceutical compounds are lipophilic in nature and have low biodegradability. These compounds enter the environment by different possible routes, such as discharge of industrially treated or partially treated wastewater, sewer lines, seepage from landfill sites, animal waste, etc. [9, 10]. Many reviews are available which describe the detrimental effects of pharmaceuticals on the ecosystem [11, 12]. Many advanced techniques like ozonation, granular activated carbon, membrane technology, and ultraviolet radiation have been used to remove pharmaceutical contaminants from water destined for human consumption. But many pharmaceutical products, for example anti-epileptics (carbamazepine, primidone), non-steroidal anti-inflammatory drugs (NSAIDs) (diclofenac, ibuprofen, indomethacin, ketoprofen), and lipid regulators (clofibric acid, gemfibrozil), are known to resist such treatment techniques, because of their high solubility and poor degradability in water. Pharmaceutical pollutants remain in the treated wastewater and return to the natural environment through various routes [13, 14]. Ibuprofen exhibited toxic effects on riverine microbial communities. Paracetamol has been found in high concentrations (up to10000 ng/L) in surface waters [15], and diclofenac has been found in the highest concentrations in the environment [16]. For all these reasons, the risk these pharmaceuticals pose to the environment can't be ruled out due to their intrinsic property of causing a biological effect [17]. Though drugs are manufactured for the well being of the community with the intention of causing health benefits, the undesirable impacts of inadvertent exposure of their by-products to the environment and ecological communities cannot be ignored. Due to effluents released by pharmaceutical industries, agricultural activities, and other industrial activities, our planet has been subjected to overexploitation, resulting in enhanced pressure on natural resources, especially the world's water reserves and hence on environmental sustainability. Therefore, it becomes absolutely essential to assess the environmental risks associated with the disposal of wastewater and treatment techniques for the subsequent return to the natural healthy environment in the most efficient way possible. Pharmaceuticals come across various biotic and abiotic agents while being in the environment and undergo several malevolent transformations. These transformations produce different metabolites and daughter compounds, which may be even more toxic than their parent compounds. Therefore, a deep understanding is required to assess and mitigate the hazards of the transformation of pharmaceuticals in the environment [18]. Many researchers have tried to develop models for the prediction of pollution levels and the corresponding risks associated with the release and accumulation of pharmaceuticals as toxicants [19, 20].

It has been found that the pharmaceutical industry generates huge quantities of toxic waste during manufacturing processes and maintenance operations. These production industries were found to be the parent sources of increased pharmaceutical concentrations in the environment than those caused by the application of drugs. Though the trace concentration of pharmaceuticals is detected in the amount of nanograms per liter, which is of a lower magnitude than the minimum therapeutic dose, it may cause considerable toxic effects on human health and aquatic environment. The United States Environmental Protection Agency has added four drugs to the most recent contaminant candidate list (CCL3) as extensively used drugs, which includes three birth control drugs and one antibiotic [21]. Because of their biological activity, there is a keen concern about their presence in the drinking water and aquatic environment. Therefore, risk assessment and management issues are gaining momentum these days. In this chapter, an overview of thc release and accumulation of pharmaceuticals and their toxic by-products in the environment is given along with the important issues related to critical risk assessment and challenges associated with their environmental exposure. The purpose of this chapter is to evaluate whether the extensive and uncontrolled use of pharmaceuticals and their unintended release into the environment may pose a potential human risk? This chapter elucidates the current and future problems associated with pharmaceutical pollution, their ecotoxicity, and the various risks posed by these micropollutants to human health and the environment, and sheds light on the need for a shift to sustainable pharmacy.

2. Occurrence and Hypostatization of Pharmaceuticals in the Environment

The discussion on the sources and origins of pharmaceuticals occurring as residues in the environment is indispensable for gaining an appreciation for the scope and magnitude of the entire issue. A lucid understanding of the origins and sources of pharmaceuticals is required to assess knowledge gaps, so that, future research and actions can be most effectively targeted. The main concern with the sources is whether they lead to immediate, delayed or direct exposures of biota without additional transport being required (via trophic food chain)? One such example is the relationship between the death of scavenging vultures in Southeast Asia and diclofenac-treated cattle [22].

Micro pollutants are commonly referred to as chemical compounds which are detected in the environment at concentrations ranging from μg/L to ng/L or even lower [23, 24]. The occurrence of pharmaceuticals as micro pollutants in water resources is an increasing safety concern for the scientific community, regulatory agencies, and the public because they could have adverse effects on the ecosystems when they are released into the aquatic environment. Easier access to emerging micro pollutants such as pharmaceuticals, pesticides, detergents, and personal care products has significantly increased the presence of such compounds in both the natural and manmade environments [25, 26]. The presence of these compounds varies from location to location in various sources of water, but concentrations are considerably higher in groundwater and surface water than in drinking water [2]. Pharmaceuticals travel a long way in the environment from the source of generation to their final disposal [27]. Figure 1 represents the pathways of pharmaceuticals from generation to final disposal. Pharmaceuticals can enter the environment by a number of distinct and varied routes. Two ways pharmaceuticals enter the environment are indirect (involuntarily) by excretion and bathing, and direct (purposefully) by disposal. Drugs that are no longer needed or desired are disposed of by throwing them in the trash or directly discarding to sewage systems.

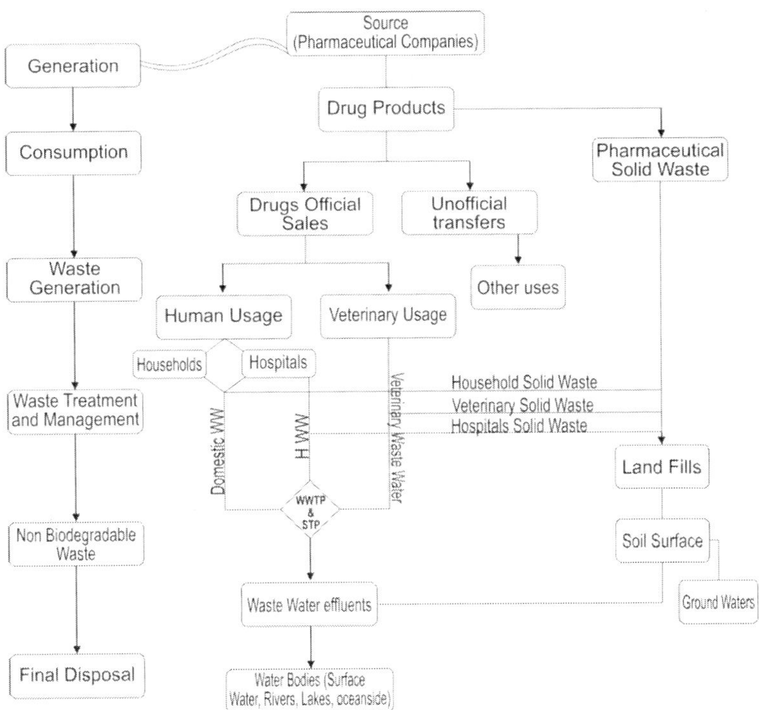

Figure 1. Different routes and possible sources of hypostatization of pharmaceuticals in environment, adopted from [51].

Table 1. Some representative classes of Pharmaceuticals reported in surface waters, wastewater, and sewage treatment plants

Pharmaceutical class	Detected compound	Use/origin
Anti-inflammatory (NSAIDS) Analgesics/ non-steroidal	Ketoprofen, naproxen, phenazone, indomethacine Acetaminophen (analgesic), diclofenac, ibuprofen	All NSAIDs have antipyretic, analgesic, and anti-inflammatory effect, NSAIDs are the most used and abused drugs in the world today.
Synthetic steroids	Diethylstilbestrol and 17-ethinylestradiol	These are synthetic steroids mainly consist of oral contraceptives as well as steroids used for substitution therapy during menopause.
Antiseptics	Triclosan and thymol	Antiseptics are chemical agents that slow or stop the growth of microorganisms (germs) on external surfaces of the body and help prevent infections.
Antibiotics/ antimicrobial	Sulfonamides, fluoroquinolones, trimetoprim, chlortetracycline, erythromycin, lincomycin, oxytetracycline, tetracycline, roxithromycin, tylosin Carbamazepine	Antiseptics (distinguished from antibiotics) destroy microorganisms inside the body and from disinfectants, which destroy microorganisms found on inanimate objects. Used for the treatment of bacterial infections in both humans and animals.
Antiepileptic	Clofibrate, bezafibrate	Antiepileptics are commonly used in medicine to stop, prevent or control seizures (convulsions, partial seizures, generalized tonic–clonic seizures, etc.).
Lipid regulators	Diazepam	Lipid regulators may be used to lower cholesterol/ triglyceride (fat-like substances) levels in the blood
Anti-anxiety	Benzodiazepines	Anti-anxiety/hypnotic agents are used to relieve anxiety, nervousness, and tension associated with anxiety disorders

Source: Adopted from [78].

Poisoning of human beings by medicines which are improperly discarded by others is a problem of increasing concern to healthcare professionals. Drug disposal is a very complex process. Previous work [28] addressed the issues and scope of drug disposal as well as other approaches to pollution reduction. The distinction between source, origin, and fate is often vague and unjustified. A number of exposures and scenarios can come into the picture in a pollutant's environmental transport chain. Any point in the chain can be considered a source, but not necessarily an origin.

Table 1 shows the uses of some pharmaceutical classes and how they are detected.

3. Active Pharmaceuticals (APIs) in the Environment

APIs are the biologically active compounds of drugs, such as tablets, capsules, creams, and injections that produce the desired effect. Basically, a pharmaceutical (in liquid or in solid form) consists of several APIs, additives, excipients, inorganic salts, and many other organic chemicals such as sugars, pigments, and dyes. A few drugs may contain endocrine disrupting chemicals, excipients, and additives. Even though APIs cover a wide range of small molecules with different biological and physiochemical properties, a small change in the structure of APIs can significantly affect the environment [29]. In this view, individual assessment for each type of API is essential in different environmental conditions. Approximately half of the 2300 APIs that are approved for human usage in Germany are considered to be potential threats to the environment because they are persistent, toxic, and bio-accumulative in nature. Consumption of medicines is an everyday occurrence on a global scale. With the increase in life expectancy, the use of pharmaceuticals is expected to increase in the future. A man of 20–24 years of age takes an average of 56 defined daily doses, while people older than 60 years need 649 defined daily doses, almost twelve times higher [30].

Most APIs remain in liquid form and, to a certain extent, pass through a wastewater treatment plant. By using special techniques like membrane filtration, activated charcoal, or ozonation; the cleaning capacity of wastewater treatment plants can be improved. Once pharmaceuticals and their metabolites reach water bodies, they may also get into groundwater, though in much lower concentrations. Traces of pharmaceuticals have also been detected in drinking water [31-35]. Pharmaceuticals and other micro-pollutants can be removed to a certain extent via physical methods, biodegradation, and chemical Advanced Oxidation Processes (AOPs) [31]. Among these technologies, ultraviolet (UV) irradiation and UV-based AOPs have become more popular in the past decade. But these micropollutants cannot be eradicated by conventional wastewater treatment because they reach water bodies through many routes.

Pharmaceuticals in the aquatic environment have been detected for more than four decades; consequently, the research work on the analysis of drugs in drinking water sources have considerably increased [36, 37]. According to research and monitoring projects by public authorities in Germany, more than 150 different APIs have been detected nationwide in different environmental matrices such as surface water. Concentrations of diclofenac, a heavily used painkiller, have been detected in the range from below 0.1 mg per litre to more than 1.0 mg per litre and show strikingly high concentrations in surface waters [38]. An analysis of fourteen pharmaceutical drugs reported some pharmaceuticals like acetaminophen (detection frequency of 0.32%), p-xanthine (0.08%), codeine (0.16%), sulfamethoxazole (0.41%), caffeine (0.24%), carbamazepine (1.5%), and trimethoprim (0.08%) have been detected at concentrations more than or equal to the detection limits of selected methods of detection. On the other hand, detection frequencies of pesticides (33%) and trihalomethanes (28%) in the same sources were found to be considerably higher [39].

4. Pharmaceutical Metabolites and Transformation Products

An understanding of pharmaceuticals' presence in the environment is a matter of keen concern. After consumption, the pharmaceuticals are metabolized in the living body to which they are administered. The degree of transformation and metabolism depends upon the chemical nature and design of the drug. Basically, two categories of metabolites exist in the environment; human metabolites and microbial metabolites. The parent compounds are released as transformed products. Ingested ibuprofen (IBF) is excreted by humans in the form of unchanged IBF, 2-4-(2 hydroxy-2-methylpropyl)-phenyl-propionic acid, and 2-4-(2-carboxypropyl)-phenyl-propionic acid at 1%, 25%, and 37%, respectively, and the remaining 14% is released as conjugated IBF [40]. About 60-70% of pharmaceuticals get metabolized by enzyme action and consumed in the human body [41], and the residual drugs get excreted in the form of human excreta. These drugs are biotransformed by microorganisms into carbon dioxide and water. However, abiotic factors such as UV, pressure, pH, and temperature change play an important role in transforming pharmaceuticals by microorganisms [42].

The presence or absence of important elements like humic acid in the wastewater matrix also affects the mechanism of degradation of pharmaceuticals. According to a study on the biodegradation of anti-inflammatory drugs, it was found that these pharmaceuticals could be transformed into many kinds of intermediate metabolites [43] as mentioned in Table 2.

Table 2. Transformation products of different categories of pharmaceutical compounds

Drug category	Parent Compound	Transformation agent	Transformation Product	Refs
Analgesics	Oxycodone Mathadone	Human metabolism	Noroxycodone 2-Ethylidine-1, 5-demethyl-3,3-diphenlypyrrolidine	[46] [47]
	Propoxyphene Tramadol	Human metabolism	Norpropoxyphene O-Desmethyltramadol	[48] [49]
Anesthetics	Ketamine	Human metabolism, photolysis	Norketamine, hydroxyl-norketamine	[50, 51]
Antacids	Cimetidine	Human metabolism	N-Glucuronide, sulfoxide, hydroxymethylcimetidine	[52]
Antibiotics	Triclosan	Chlorine	Tri & tetra chlorinated dibenzo-p-dioxins	[53]
	Amoxicillin	Functionalised bentonite	Amoxicillic acid	[54]
	Erythromycin	Biodegradation sludge	L-Cladinose and D-desosamine	[55]
	Chloramphenicol	Human metabolism	Glucuronide conjugate	[6]
Beta blockers	Acebutalol	Chlorine	Dealkylation, hydroxylation, chlorination, oxidation products	[56]
Diuretic	Furosemide	Photolysis	FUR 1-5 (furanylmethyl +/-, hydroxylation, dehalogenation)	[57]
Hypertension	Irbesartan Valsartan	Waste water treatment plant	IRB_444, IRB_442, IRB_446	[58]
Lipid regulators	Atorvastatin	Sunlight	5 TPs-pyrrole ring, alkyl/aryl shift, lactum ring. Lactonization and cyclization	[59]
NSAIDS	Diclofenac	Oxidation, esterification & demethylation Human metabolism	DIC 1-6, glocuronide, sulphate conjugates	[48]
	Ibuprofen	Oxidation, esterification & demethylation Human metabolism	(p)-2-40-(2-hydroxy-2-methylpropyl)-phenylpropionic acid (25%) &(p)-2-40-(2-carboxypropyl)-phenylpropionic acid (37%) conjugated ibuprofen (14%)	[48, 60]
	Ketoprofen	UV free radical, dechlorination, oxidation	KET 1-8, 2-(3-benzoylphenyl)-propanoic acid & glucuronides	[60]
Plasticizers	Bisphenol-A	Ozonation	Atechol, orthoquinone, muconic acid derivatives of BPA, benzoquinone & 2-(4-hydroxyphenly)-propan-2-ol	[61]

For example, among the three intermediate metabolites (hydroxyl-IBF, carboxy-IBF, and carboxy-hydratropic acid), the hydroxyl-IBF was the primary product formed during aerobic biotransformation, whereas the carboxy-hydratropic acid was formed during anaerobic conditions. Carboxy-IBF accounted for less than 10% of the transformation products and was the minor product. This study indicates that in order to obtain controlled outcomes and safer transformation products, the biodegradation reaction can be controlled [44]. Transformations of pharmaceuticals may occur through many pathways like dehalogenation, isomerisation, hydroxylation, alkylation, decarboxylation, ring formation, and many other chemical transformations [45]. An emphasis is given on biotransformation and photo-transformation products of pharmaceuticals formed in the presence of such compounds in the sewage treatment plants and the aquatic environment along with the challenges related to their ecological risk.

5. Ecological and Human Risk Assessment: Trends and Future Requirements

It has been recognized globally that the human body and domestic animals release active pharmaceutical ingredients which are found in treated and untreated surface water, sewage effluents, ground water, and drinking water [62]. Several therapeutic classes have been found in the environment, such as anti-inflammatory drugs, contraceptive pills, antineoplastics, beta-blockers, antiseptics, X-ray contrast agents, etc. For example, since pharmaceutical products are biologically active, such contamination must be assessed in connection with the possible human, ecological, and environmental risks. For example, endocrine disruption in fish is caused by very low levels (ng L^{-1}) of exposure to synthetic hormones [63]. Many other studies have demonstrated that wastewater effluents can cause reproductive abnormalities in fish. Moreover, the exposure of fish to trace concentrations of birth control compounds found in the environment shows a considerable decrease in reproductive success; impacts on fish populations are also possible [12]. Risk assessments for possible undesirable effects in human beings, particularly for sub-therapeutic concentrations and persistent exposure, are scarcely available. These assessments have become a global scientific priority for the "International Decade for Action: Water for Life," launched in 2005 and the Millennium Development Goals with the objective of doubling the world's population with sustainable access to drinking water and sanitation by 2015

[64]. Many reviews are available explaining the detrimental effects of pharmaceutical compounds on the ecosystem [65]. The same is explained with the aid of Figure 2.

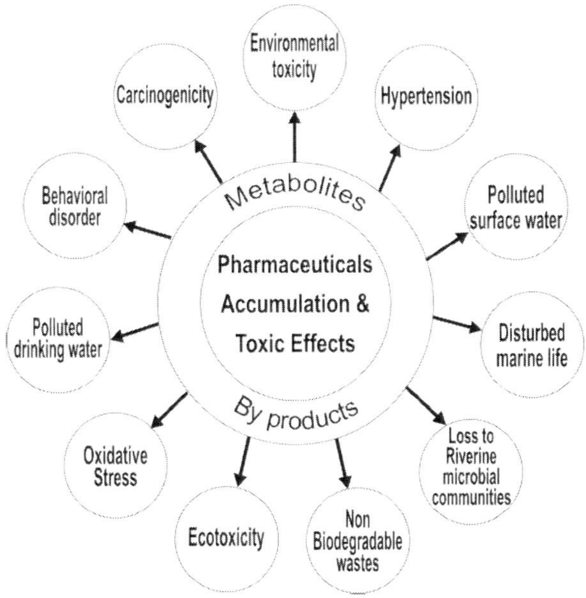

Figure 2. Possible toxic effects of accumulation of pharmaceuticals on human health and environment.

In general, not much information is available concerning the effects of active substances on terrestrial and aquatic organisms, but in long-term tests, the effects on fish, algae, bacteria, and daphnia have been demonstrated using low concentrations [66, 67]. Acute toxicity is assessed by evaluating the effect of a toxicant on the growth of an organism used as an indicator. *Daphnia magna* (a small planktonic crustacean), *Pseudokirchneriella, Brachionus-calyciflorus,* etc. are other indicators on which ecotoxicity studies are done [68, 69]. For biological assay, a characteristic model organism needs to be selected on the basis of morphological, physiological, genetic, molecular, and ecological traits. This model organism should be fast-growing, reproducible, and highly typified. *Zebrafish* has been used as a model organism to study human pathology because it contains 70% of human coding genes for diseases [70]. The effective concentration of diclofenac for chronic toxicity was in the range of wastewater concentrations [71, 72] whereas the concentration of propranolol for zooplankton and benthic organisms was in close proximity to the maximally measured sewage treatment plant effluent concentrations [73].

Advanced technologies (ozonation, membrane technology, UV radiation, granular activated carbon, etc.) have been used for the removal of environmental contaminants from the water meant for human usage. But with such treatment techniques, many types of drugs like NSAIDS, anti-epileptics, and lipid regulators cannot be removed due to their high solubility and poor degradability in water. Consequently, water resources have shown higher concentration levels of pharmaceuticals [74]. Although, degradation of pharmaceuticals in the environment can be done by biotic or abiotic methods [75, 76], even then, they may persist because of continual infusion into aquatic media from STP effluents [77, 78]. For this very reason, and due to the intrinsic property of pharmaceuticals to cause a biological consequence, the threat and risk they pose to the environment cannot be ruled out [20]. The multiple threats to the environment, humans, and ecosystems are mentioned in Figure 2. Therefore, there is a need to consider and examine the effects of various drugs affecting the same receptor. A recent study revealed that current levels of exposure to organophosphate pesticides slow down the metabolism of the antidepressant imipramine by human recombinant enzymes and in liver microsomes [79]. The risk assessment of mixtures and the interaction potential between pharmaceutical products and other environmental contaminants need to be studied thoroughly. The adverse effects of a wide range of pharmaceutical products, pesticides, and other environmental contaminants (as potential inhibitors or inducers of drug-metabolizing enzymes) on human health must be thoroughly studied.

Combining toxicological data with toxicogenomics data can provide substantial information for assessing pharmaceutical products. For example, quantitative structure-activity relationships (QSARs) can correlate the activity of chemicals with their structure through the use of statistical tools. These approaches can provide unparalleled results for predicting the activities of those pharmaceutical products that have not been tested yet. In conjunction with toxicity, the assessment of persistence is of utmost importance for environmental significance. The organic pollutants, if persistent for a long time, may produce varied effects, and longer exposure may result in multidimensional contamination of the ecosystem and environment. These complications cannot be determined in advance with the currently available testing methods [80].

6. Possible Measures, Activities, and Future Approaches to Reduce the Input of Pharmaceuticals into the Environment

With the understanding of the sources or origin and accumulation of pharmaceuticals as pollutants, those sources most amenable to lessening their connections with the environment can be identified. Pollution minimization encompasses a wide range of actions, including reduced dosage, waste treatment, waste/contaminant storage, recycling/reuse, disposal, and pollution prevention. A wide range of techniques, actions, and activities could be designed and implemented to reduce the environmental residues contributed by pharmaceuticals. Despite an appropriate legislative framework to check the inflow of pharmaceuticals into drinking water, especially in advanced nations, the problem still persists [81]. A number of conventional wastewater treatment technologies have been reported in the available literature. However, these technologies are not sufficient and completely effective in the removal of pharmaceuticals from contaminated water. For example, physicochemical treatments such as activated carbon adsorption, reverse osmosis, nano-filtration, coagulation–flocculation, and membrane filtration [82, 83, 84] were discovered but could not eradicate pharmaceuticals and other pollutants from the environment. Instead, they generate other toxic by-products as the end products.

A literature survey revealed that treatment of wastewater containing organic pollutants is a complicated, expensive, and time-consuming process, and most wastewater treatment plants are not designed for emergent pollutants' removal. Therefore, a more realistic, practical, versatile, and economically viable technique is needed at this hour. However, AOPs in conjunction with other conventional treatment technologies have been implemented to remove pharmaceuticals and endocrine disruptors from wastewater. But high operational and maintenance costs, complicated treatment procedures [85], non-availability of chemicals (O_3, H_2O_2, F_2+), generation of toxic residues and other factors need to be considered while implementing the AOPs. AOPs when combined with biological processes are not often effective against the targeted pollutant degradation due to the inhibitory effect of hydroxyl radicals by the biomass. Although free radicals are short-lived and rapid in their work without leaving long-term residues if applied properly, they must be managed well so as not to leave partially degraded secondary pollutants. Therefore, physical treatment such as filtration can be employed before or after the AOPs. Thus, the application of physical methods along with biological methods may help to remove large insoluble particles, reducing turbidity and interfering ions from the aqueous matrix. Once this is done, the degradation efficiency of AOPs will be improved.

The biological treatment approach (specifically the activated sludge approach), which was originally designed to remove organic carbon, is the most commonly used technology for waste water treatment. The use of this technology is further extended to remove nitrogen and phosphorus also. Many studies have reported that removal of organic pollutants takes place in biological treatment systems by the combined treatment through biosorption and biodegradation. Prediction of the removal efficiency of micropollutants is difficult due to their diverse chemistry and heterogeneity (they may be classified as easily, moderately or poorly degradable). Complex microbial consortia can be developed to enhance the degradation of micropollutants by biological method because oxidative enzymes as biocatalysts have great potential for organic waste breakdown. The cytochrome P450 (an efficient group of monooxygenases) is highly effective in the degradation of drugs and toxins in microorganisms. A class of copper-containing oxidase enzyme (laccases) is also used by microorganisms to break down lignin [86]. Both these enzymes effectively degrade a wide range of organic micropollutants but are novices in wastewater treatment. Biological treatment systems have a wide variety of active microorganisms that can be used to stimulate a wide range of chemical reactions. Therefore, use of biological treatment systems on a pilot scale is required to test their efficacy for wastewater treatment containing pharmaceutical pollutants.

7. Move towards Green and Sustainable Pharmacy

On the basis of the above discussion, it is hard to ignore that pharmaceuticals are one of the biggest contributors to environmental pollution. As one of the biggest polluters, the pharmaceutical industry has a responsibility to be more eco-friendly. Green pharmacy includes the design of pharmaceutical products and processes that significantly reduce the generation of toxic and hazardous end products to ensure environmental security and health [87]. Another step that pharmaceutical industries can take to minimize the environmental impact of pharmaceutical products is the collection of unused and expired medicines from consumers. In the document called "Green Pharmacy Practice," the International Pharmaceutical Federation makes recommendations to community pharmacies regarding pharmaceutical waste collection and disposal. Pharmacists can communicate with the users and provide information regarding medicinal products, including the environmental risks associated with them. A pharmaceutical can be green in terms of waste generated during its synthesis, but it may turn toxic due to accumulation in the environment post excretion. Similarly, green chemicals and pharmaceuticals, if produced in high volumes, may become challenging and unsustainable [87].

Since the pharmaceutical industry produces a significant amount of waste and greenhouse gases apart from utilizing great volumes of water, there is a need to implement sustainability in pharmaceutical production. High energy consumption, diversified supply chains, and environmental pollution are some of the important factors that need to be addressed globally. By improving production processes and optimizing supply chains, the industry can enhance value and also ensure environmental sustainability. The industry can ensure social sustainability by focusing on providing access to affordable medicines and treatments to developing countries with poor and underfunded healthcare systems. Benign production is the building block of green and sustainable pharmacy, with a paradigm shift in the future.

Conclusion

Conclusively, the presence of pharmaceuticals in the global environment in the form of micropollutants can't be denied so far, and the risk analysis is characterized by its inherent multidisciplinary nature. The long-term impacts of drugs occur on environmental ecosystems, humans, and biota. The direct link between environmental, human, and ecological risk assessment is transparent due to the direct and indirect effects of chemical species on the environment and human health. Toxicokinetics and toxicodynamics can define the pathways of metabolism for even single pharmaceutical products and mixtures. Many organic pollutants are present in wastewater. The removal of pollutants is highly recommended to obtain good quality water. Current systems of wastewater treatment are not designed to degrade organic micropollutants completely. The physiochemical treatments are effective but quite expensive, and the biological treatments are not so effective and are non-consistent. The next generation of biological waste water treatment could be based on the synthesis of enzymes which can degrade micropollutants and can discover and employ the metabolic traits of unconventional microbes. New tools, such as QSAR modeling, genomic sciences, and probabilistic approaches, have the potential to assist scientists in qualifying and quantifying biological effects of pharmaceutical products. The approach to "International Chemical Management" (2006) assumes a predominant role in promoting a policy framework to improve and enhance chemical safety around the world.

References

[1] Xagoraraki, I., & Kuo, D. (2008). "Water pollution: Emerging contaminants associated with drinking water." *In International Encyclopedia of Public Health,* edited byK. Heggenhougen, S. Quah, 539-550. Academic Press: San Diego, CA, USA. doi: 10.1016/B978-012373960-5.00292-6.

[2] Benotti, M.J., Trenholm, B.A., Vanderford, B.J., Holady, J.C., Stanford, B.D., & Snyder, S.A. (2009). "Pharmaceuticals and endocrine disrupting compounds in U.S. drinking water." *Environ. Sci. Technol,* 43, 3 : 597-603. doi: 10.1021/es801845a.

[3] Stackelberg, P.E., Gibs, J., Furlong, E.T., Meyer, M.T., Zaugg, S.D., & Lippincott, R.L. (2007). "Efficiency of conventional drinking-water-treatment processes in removal of pharmaceuticals and other organic compounds." *Sci. Total Environ.,* 377: 255-272. doi: 10.1016/j.scitotenv.2007.01.095.

[4] Schwab, R.W., Hayes, E.P., Fiori, J.M., Mastrocco, F.J., Roden, N.M., Cragin, D., Meyerhoff, R.D., D'Aco, V.J., & Anderson, P.D. 2005. "Human pharmaceuticals in US surface waters: A human health risk assessment." *Regul. Toxicol. Pharmacol,* 42 : 296-312. doi: 10.1016/j.yrtph.2005.05.005.

[5] Nikolaou, A., Meric, S., & Fatta, D. (2007). "Occurrence patterns of pharmaceuticals in water and wastewater environments" *Anal. Bioanal. Chem.,* 387,4: 1225-1234. doi: 10.1007/s00216-006-1035-8.

[6] Archer, E., Petrie, B., Kasprzyk-Hordern, B., & Wolaardt, G.M. (2017). "The fate of pharmaceuticals and personal care products (PPCPs), endocrine disrupting contaminants (EDCs), metabolites and illicit drugs in WWTW and environmental waters." *Chemosphere,* 174 : 437-446. https://doi.org/10.1016/j.chemosphere.2017.01.101.

[7] Behera, S.K., Kim, H.W., Oh, J.E., & Park, H.S. (2011). "Occurrence and removal of antibiotics, hormones and several other pharmaceuticals in wastewater treatment plants of the largest industrial city of Korea." *Sci. Total Environ.,* 409, 20 : 4351-4360. doi: 10.1016/j.scitotenv.2011.07.015.

[8] Carlsson, C., Johansson, A.K., Alvan, G., Bergman, K., & Kuhler, T. (2006). "Are pharmaceuticals potent environmental pollutants? Part I: environmental risk assessments of selected active pharmaceutical ingredients." *Sci. Total Environ.,* 364, 1-3 : 67-87. doi: 10.1016/j.scitotenv.2005.06.035.

[9] Larsson, D.G.J., de Pedro, C., & Paxeus, N. (2007). "Effluent from drug manufactures contains extremely high levels of pharmaceuticals." *J. Hazard Mater,* 148 : 751-75. doi: 10.1016/j.jhazmat.2007.07.008.

[10] Potera, C. (2000). "Drugged Drinking Water" *Environ Health Perspect.,*108, : A446. https://doi.org/10.1289/ehp.108-a446.

[11] Dorne, J.L.C.M, Ragas, A.M.J., Frampton, G.K., Spurgeon, D.S., & Lewis, D.F. (2007) "Trends in Human Risk Assessment of Pharmaceuticals." *Anal. Bioanal. Chem,* 387 : 1167-1172. doi: 10.1007/s00216-006-0961-9.

[12] Fent, K., Weston, A.A., & Caminada, D. (2006). "Ecotoxicology of human pharmaceuticals" *Aquat. Toxicol,* 76, 2: 122– 159. https://doi.org/10.1016/j.aquatox. 2005.09.009.

[13] Lee, S.-H., Kim, K.-H., Lee, M., & Lee, B.-D. (2019). "Detection status and removal characteristics of pharmaceuticals in wastewater treatment effluent." *J. Water Process. Eng.,* 31 : 100828. doi: 10.1016/j.jwpe.2019.100828.

[14] Brillas, E. (2020). "A review on the photoelectro-Fenton process as efficient electrochemical advanced oxidation for wastewater remediation. Treatment with UV light, sunlight, and coupling with conventional and other photo-assisted advanced technologies." *Chemosphere,* 250 : 126198. doi: 10.1016/j.chemosphere. 2020.126198.

[15] Boxall, A.B., Sinclairm C.J., Fenner, K., Kolpin, D., & Maund, S. J. (2004). "When synthetic chemicals degrade in the environment." *Environ Sci Technol,* 38 : 368A–375A. doi: 10.1021/es040624v.

[16] Jones, O.A., Lester, J.N., & Voulvoulis, N. (2005). "Pharmaceuticals : A Threat to Drinking Water." *Trends Biotechnol,* 23, 4 : 163–167. https://doi.org/10.1016/ j.tibtech.2005.02.001.

[17] Henschel, K.P., Wenzel, A., Diedrich, M., & Fliedner, A. (1997). "Environmental hazard assessment of pharmaceuticals." *Regul Toxicol Pharmacol,* 25 : 220–225. doi: 10.1006/rtph.1997.1102.

[18] Maldonado-Torres, S., Gurung, R., Rijal, H., Chan, A., Acharya, S., Rogelj, S., Piyasena, M., & Rubasinghege, G. (2018). "Fate, transformation, and toxicological impacts of pharmaceutical and personal care products in surface waters." *Environ. Health Insights,* 12. https://doi.org//10.1177/1178630218795836.

[19] Tropsha, A., & Golbraikh, A. (2007). "Predictive QSAR modeling workflow, model applicability domains, and virtual screening." *Curr. Pharm. Des.,* 13, 34: 3494-3504. doi: 10.2174/138161207782794257.

[20] Tropsha, A. (2010). "Best practices for QSAR model development, validation, and exploitation." *Mol. Inform.,* 29, 6-7: 476-488. https://doi.org/10.1002/minf.201000 061.

[21] FEPA (Federal Environment Protection Agency). (1988). "Guideline and Standards for Environmental Pollution Control in Nigeria." *Federal Government Press,* Lagos. 238-291. https://www.scirp.org/(S(lz5mqp453edsnp55rrgjct55.))/reference/ referencespapers.aspx? referenceid=2321128.

[22] Oaks, J.L., Gilbert, M., Virani, M.Z., Watson, R.T., Meteyer, C.U., Rideout, B.A., Shivaprasad, H.L., Ahmed, S., Chaudhry, M.J.I., Arshad, M., Mahmood, S., Ali A., & Khan. A. A. (2004). "Diclofenac residues as the cause of vulture population decline in Pakistan." *Nature,* 427 : 630–633. doi: 10.1038/nature02317.

[23] Hollender, J., Singer, H., & Mcardell, C. (2008). "Polar organic micropollutants in the water cycle." In *Dangerous pollutants (Xenobiotics) in urban water cycle.* Edited by Hlavinek, P., Bonacci, O., Marsalek, J., & Mahrikovaa, I. 103–116.

Berlin: Springer. https://www.springerprofessional.de/en/dangerous-pollutants-xeno biotics-in-urban-water-cycle/1607154.

[24] "Strategy for micro-pollutants – strategy for municipal and industrial wastewater." (2010). In *IKSR-CIPR-ICBR. International Commission for the Protection of the Rhine (ICPR)*: 181 https://www.iksr.org/fileadmin/user_upload/DKDM/Dokumente/Fachberichte/EN/rp_En_0181.pdf.

[25] Bunke, D., Moritz, S., Brack, W., HERRaÉz, D.L., Posthuma, L., & Nuss, M. (2019). "Developments in society and implications for emerging pollutants in the aquatic environment." *Environ. Sci. Europe,* 31, 32. https://doi.org/10.1186/s12302-019-0213-1.

[26] Wanda, E., Nyoni, H., Mamba, B., & Msagati, T., (2017). "Occurrence of emerging micropollutants in water systems in Gauteng, Mpumalanga, and North West Provinces, South Africa." *Int. J. Environ. Res. public. Health,* 14, 1: 79. doi: 10.3390/ijerph14010079.

[27] Wilkinson, J.L., Hooda, P.S., Swinden, J., Barker, J., & Barton S. (2018). "Spatial (bio)accumulation of pharmaceuticals, illicit drugs, plasticisers, perfluorinated compounds and metabolites in river sediment, aquatic plants and benthic organisms." *Environ. Pollut.,* 234: 864-875. doi: 10.1016/j.envpol.2017.11.090.

[28] Daughton, C.G. (2003). "Cradle-to-Cradle Stewardship of Drugs for Minimizing Their Environmental Disposition While Promoting Human Health. II. Drug Disposal, Waste Reduction, and Future Directions." *Environ. Health Perspect,* 111: 775–785. https://ehp.niehs.nih.gov/doi/epdf/10.1289/ehp.5948.

[29] Cunningham, V. (2008). "Special characteristics of pharmaceuticals related to environmental fate." *ChemInform,* 36, 46:23-34. doi: 10.1007/978-3-540-74664-5_2.

[30] Klatte, S., Schaefer, H.C., & Hempel, M. (2017). "Pharmaceuticals in the environment – A short review on options to minimize the exposure of humans, animals and ecosystems." *Sustainable Chemistry and Pharmacy,* 5 : 61-65. https://agris.fao.org/agris-search/search.do?recordID=US201800374101.

[31] Liu, Z. H., Kanjo, Y., & Mizutani, S. (2009). "Removal mechanisms for endocrine disrupting compounds (EDCs) in wastewater treatment – physical means, biodegradation, and chemical advanced oxidation: A review." *Science of the Total Environment,* 407, 2 : 731-748. https://doi.org/10.1016/j.scitotenv.2008.08.039.

[32] Rabiet M., Togola, A., Brissaud, F., Seidel, J.L., Budzinski, H., & Poulichet, F.E. (2006). "Consequences of treated water recycling as regards pharmaceuticals and drugs in surface and ground waters of a medium sized Mediterranean Catchment." *Environmental Science & Technology,* 40, 17: 5282-5288. doi: 10.1021/es060528p.

[33] Kleywegt S., Pileggi, V., Yang, P., Hao, C., Xhao, X., Rocks, C., Thach, S., Cheung, P., & Whitehead, B. (2011). "Pharmaceuticals, hormones and bisphenols A in untreated source and finished water in Ontario, Canada: Occurrence and Treatment Efficiency." *Science of the Total Environment,* 409, 8: 1481-1488. doi: 10.1016/j.scitotenv.2011.01.010.

[34] Togola, A., & Budzinski, H. (2008). "Multi-residue analysis of pharmaceutical compounds in aqueous samples". *Journal of Chromatography A,* 1177, 1: 150-158. doi: 10.1016/j.chroma.2007.10.105.

[35] Boleda, MR., Galceran, M.T., & Ventura, F. (2011). "Behavior of Pharmaceuticals and drugs of abuse in a drinking water treatment plant (DWTP) using combined conventional and ultrafiltration and reverse osmosis (UF/RO) treatments." *Environmental Pollution,* 159, 6: 1584-1591. doi: 10.1016/j.envpol.2011.02.051.

[36] Wu, M., & Janssen, S. (2011). "Dosed without prescription: A framework of preventing pharmaceutical contamination of our nation's drinking water." *Environ SciTechnol*, 45, 2: 366-7. doi: 10.1021/es104032a.

[37] Bruce, G.M., Pleus, R.C., & Snyder, S.A. (2010). "Toxicological Relevance of Pharmaceuticals in Drinking Water." *Environment Science Technology,* 44, 14: 5619-5626. doi: 10.1021/es1004895.

[38] Weber, F.A., aus der Beek, T., Bergmann, A., Carius, A., Grüttner, G., Hickmann, S., Ebert, I., Hein, A., Kuester, A., Rose, J., Koch-Jugl, J., & Stolzenberg, H.C. (2014). *Pharmaceuticals in the environment – the global perspective.* German Environ Agency. https://www.pharmaceuticals-in-the-environment.org. Date of last access: (04.04.16.)

[39] Fram, M.S., & Belitz, K. (2011). "Occurrence and concentrations of pharmaceutical compounds in groundwater used for public drinking-water supply in California." *Science of the Total Environment,* 409: 3409-3417. https://doi.org/10.1016/j.scitotenv.2011.05.053.

[40] Kasprzyk-Hordern, B. (2010). "Pharmacologically active compounds in the environment and their chirality," *Chem. Soc. Rev,* 39, 11: 4466-4503. doi: https://doi.org/10.1039/C000408C.

[41] Escher, B.I., Baumgartner, R., Koller, M., Treyer, K., Lienert, J., & McArdell, C.S. (2011). "Environmental toxicology and risk assessment of pharmaceuticals from hospital wastewater." *Water Res.*, 45, 1: 75-92. https://doi.org/10.1016/j.watres.2010.08.019.

[42] Stuart, M.E., & Lapworth, D.J. (2014). "Transformation products of emerging organic compounds as future groundwater and drinking water contaminants." in *Transformation Products of Emerging Contaminants in the Environment: Analysis, Processes, Occurrence, Effects and Risks.* Edited by Dimitra A., Lambropoulou, Leo M. L. Nollet. 65-86. Wiley Online Library. https://doi.org/10.1002/9781118339558. ch02.

[43] Quintana, J.B., Weiss, S., & Reemtsma, T. (2005). "Pathways and metabolites of microbial degradation of selected acidic pharmaceuticals and their occurrence in municipal wastewater treated by a membrane bioreactor." *Water Res.*, 39, 12 : 2654-2664. doi: 10.1016/j.watres.2005.04.068.

[44] Ram S.K., Panidepu H., Cheernam V., & Tyagi, R.D. (2020). "Pharmaceutical metabolites and their by-products in hospital wastewater." in *Current Developments in Biotechnology and Bioengineering*: 43-78. doi: 10.1016/B978-0-12-819722-6.00002-X.

[45] Terzic, S., Senta, I., Matosic, M., & Ahel, M. (2011). "Identification of biotransformation products of macrolide and fluoroquinolone antimicrobials in membrane bioreactor treatment by ultrahigh-performance liquid chromatography/quadrupole time-of-flight mass spectrometry." *Anal. Bioanal. Chem.*, 401, 1: 353-363. doi: 10.1007/s00216-011-5060-x.

[46] Yargeau, V., Taylor, B., Li, H., Rodayan, A., & Metcalfe, C.D. (2014). "Analysis of drugs of abuse in wastewater from two Canadian cities." *Sci. Total Environ.*, 487: 722-730. https://doi.org/10.1016/j.scitotenv.2013.11.094.

[47] Mackuľak, T., Mosný, M., Grabic, R., Golovko, O., Koba, O., & Birosová, L. (2015). "Fenton-like reaction: A possible ˇ way to efficiently remove illicit drugs and pharmaceuticals from wastewater." *Environ. Toxicol. Pharmacol.*, 39, 2 : 483-488. doi: 10.1016/j.etap.2014.12.016.

[48] Petrie, B., Barden, R., & Kasprzyk-Hordern, B. (2015). "A review on emerging contaminants in wastewaters and the environment: current knowledge, understudied areas and recommendations for future monitoring." *Water Res.*, 72: 3-27. https://doi.org/10.1016/j.watres.2014.08.053.

[49] Rúa-Gómez, P.C., & Püttmann, W. (2012). "Impact of wastewater treatment plant discharge of lidocaine, tramadol, venlafaxine and their metabolites on the quality of surface waters and groundwater." *J. Environ. Monit.*, 14, 5: 1391-1399. https://doi.org/10.1039/C2EM10950F.

[50] Castiglioni, S., Borsotti, A., Senta, I., & Zuccato, E. (2015). "Wastewater analysis to monitor spatial and temporal patterns of use of two synthetic recreational drugs, ketamine and mephedrone, in Italy." *Environ. Sci. Technol*, 49: 5563-5570. doi: 10.1021/es5060429.

[51] Lin, A.Y.C., Lee, W. N., & Wang, X. H. (2014). "Ketamine and the metabolite norketamine: persistence and phototransformation toxicity in hospital wastewater and surface water." *Water Res.*, 53: 351-360. doi: 10.1016/j.watres.2014.01.022.

[52] Huang, G. Y., Liu, Y. S., Chen, X. W., Liang, Y. Q., Liu, S. S., Yang, Y. Y., Hu, L. X., Shi, W. J., Tian, F., Zhao, J.L., Chan, J., & Yang, G.-G. (2016). "Feminization and masculinization of western mosquitofish (*Gambusia affinis*) observed in rivers impacted by municipal wastewaters." *Sci. Rep.*, 6: 20884. doi: 10.1038/srep20884.

[53] Buth, J.M., Ross, M.R., McNeill, K., & Arnold, W.A. (2011). "Removal and formation of chlorinated triclosan derivatives in wastewater treatment plants using chlorine and UV disinfection." *Chemosphere,* 84, 9: 1238-1243. doi: 10.1016/j.chemosphere.2011.09.003.

[54] Zha, S., Zhou, Y., Jin, X., & Chen, Z. (2013). "The removal of amoxicillin from wastewater using organobentonite." *J. Environ. Manage.*, 129: 569-576. doi: 10.1016/j.jenvman.2013.08.032.

[55] Wei, L., Qin, K., Zhao, N., Noguera, D. R., Qiu, W., Zhao, Q., Kong, X., Xhang, W., & Kabutey, F.T. (2017). "Transformation of erythromycin during secondary effluent soil aquifer recharging: removal contribution and degradation path." *J. Environ. Sci.*, 51: 173-180. doi: 10.1016/j.jes.2016.08.004.

[56] Khalit, W.N.A.W., & Tay, K.S. (2016). "Aqueous chlorination of acebutolol: kinetics, transformation by-products and mechanism." *Environ. Sci. Pollut. Res.*, 23, 3: 2521-2529. doi: 10.1007/s11356-015-5470-y.

[57] Jakimska, A., Sliwka-Kaszyńska, M., Reszczyńska, J., Namieśnik, J., & Kot-Wasik, A. (2014). "Elucidation of transformation pathway of ketoprofen, ibuprofen, and furosemide in surface water and their occurrence in the aqueous environment using UHPLC-QTOF-MS." *Anal. Bioanal. Chem.*, 406, 15: 3667-3680. doi: 10.1007/s00216-014-7614-1.

[58] Letzel, T., Bayer, A., Schulz, W., Heermann, A., Lucke, T., Greco, G., Grosse, S., Schussler, W., Sengl, M., & Letzel, M. (2015). "LCMS screening techniques for wastewater analysis and analytical data handling strategies: sartans and their transformation products as an example." *Chemosphere,* 137: 198-206. doi: 10.1016/j.chemosphere.2015.06.083.

[59] Schulz, M., Löffler, D., Wagner, M., & Ternes, T. A. (2008). "Transformation of the X-ray contrast medium iopromide in soil and biological wastewater treatment." *Environ. Sci. Technol.,* 42, 19: 7207-7217. https://www.eu-neptune.org/Publications%20and%20Presentations/Schulz_2008.pdf.

[60] Li, J., Ma, L. Y., Li, L. S, & Xu, L. (2017). "Photodegradation kinetics, transformation, and toxicity prediction of ketoprofen, carprofen, and diclofenac acid in aqueous solutions." *Environ. Toxicol. Chem.*, 36: 3232-3239. doi: 10.1002/etc.3915.

[61] Deborde, M., Rabouan, S., Mazellier, P., Duguet, J.-P., & Legube, B. (2008). "Oxidation of bisphenol A by ozone in aqueous solution" *Water Res.,* 42, 16: 4299-4308. doi: 10.1016/j.watres.2008.07.015.

[62] Seiler, J. P. (2002). "Pharmacodynamic activity of drugs and ecotoxicology : can the two be connected?" *Toxicol Lett.,* 131, 1-2: 105-115. https://doi.org/10.1016/S0378-4274(02)00045-0.

[63] Schulman, L. J., Sargent, E.V., Naumann, B. D., Faria, E. C., Dolan, D. G., & Wargo, J. P. (2002). "A Human Health Risk Assessment of Pharmaceuticals in the Environment." *Human & Ecol Risk Assess*, 8: 657-680. https://doi.org/10.1080/20028091057141.

[64] Fent, K., Weston, A.A., & Caminada, D. (2006). "Ecotoxicology of human pharmaceuticals." *Aquat. Toxicol.*, 76: 122– 159. https://doi.org/10.1016/j.aquatox.2005.09.009.

[65] Crane, M., Watts, C., & Boucard, T. (2006). "Chronic aquatic environmental risks from exposure to human pharmaceuticals." *Sci Total Environ.*, 367: 23-41. https://doi.org/10.1016/j.scitotenv.2006.04.010.

[66] Holten-Lutzhøft, H.C., Halling-Sørensen, B., & Jørgensen, S.E. (1999). "Algal toxicity of antibacterial agents applied in Danish fish farming." *Arch. Environ. Contam. Toxicol.*, 36:1–6. doi: 10.1007/S002449900435.

[67] Yamashita, N., Yasojima, M., Nakada, N., Miyajima, K., Komori, K., Suzuki, Y., & Tanaka, H. (2006). "Effects of antibacterial agents, levofloxacin and clarithromycin, on aquatic organisms." *Water Sci. Technol,* 53: 65–72. doi: 10.2166/wst.2006.338.

[68] Sanderson, H., Johnson, D.J., Wilson, C.J., Brain, R.A., & Solomon, K.R. (2003). "Probabilistic hazard assessment of environmentally occurring pharmaceuticals toxicity to fish, daphnids and algae by ECOSAR screening." *Toxicol. Lett.,* 144, 3: 383-395. doi: 10.1016/S0378-4274(03)00257-1.

[69] Kar, S., & Roy, K. (2012). "Risk assessment for the ecotoxicity of pharmaceuticals an emerging issue." *Expert Opin. Drug Saf.,* 11: 235-274. doi: 10.1517/14740338.2012.644272.

[70] Ankeny, R.A., & Leonelli, S. (2011). "What's so special about model organisms?" *Stud. Hist. Philos. Sci. Part A,* 42, 2: 313-323. doi: 10.1016/j.shpsa.2010.11.039.

[71] Triebskorn, R., Casper, H., Heyd, A., Eikemper, R., Kohler, H.R., & Schwaiger, J. (2004). "Toxic effects of the non-steroidal anti-inflammatory drug diclofenac. Part II: cytological effects in liver, kidney, gills and intestine of rainbow trout (*Oncorhynchus mykiss*)." *Aquat. Toxicol.,* 68: 151–66. doi: 10.1016/j.aquatox.2004.03.015.

[72] Triebskorn, R., Casper, H., Scheil, V., & Schwaiger, J. (2005). "Ultrastructural effects of pharmaceuticals (carbamazepine, clofibric acid, metoprolol, diclofenac) in rainbow trout (*Oncorhynchus mykiss*) and common carp (*Cyprinus carpio*)." *Aquat. Toxicol.,* 75: 53–64. doi: 10.1007/s00216-006-1033-x.

[73] Ternes, T.A., Meisenheimer, M., McDowell, D., Sacher, F., Brauch, H.J., Heist-Guide, B., Preuss, G., Wilme, U., & Zulei-Seibert, N. (2002). "Removal of pharmaceuticals during drinking water treatment." *Environ Sci Technol,* 36: 3855–3863. doi: 10.1021/es015757k.

[74] Halling-Sorensen, B., Nielsen, S.N., Lanzky, P.F., Ingerslev, F., HoltenLützhoft, HC., & Jorgensen, S.E. (1998). "Occurrence, fate and effects of pharmaceutical substances in the environment-A review." *Chemosphere,* 36: 357–393. https://doi.org/10.1016/S0045-6535(97)00354-8.

[75] Andreozzi, R., Marotta, R., Pinto, G., & Pollio, A. (2002). "Carbamazepine in water: Persistence in the environment, ozonation treatment, and preliminary assessment on algal toxicity." *Water Res.,* 36: 2869– 2877. doi: 10.1016/S0043-1354(01)00500-0.

[76] Tixier, C., Singer, H.P., Oellers, S., & Muller, S.R. (2003). "Occurrence and fate of carbamazepine, clofibric acid, diclofenac, ibuprofen, ketoprofen, and naproxen in surface waters." *Environ Sci Technol,* 37: 1061–1068. doi: 10.1021/es025834r.

[77] Daughton, C., &Ternes, T. A. (1999). "Pharmaceuticals and personal care products in the environment: Agents of subtle change?" *Environ Health Perspect.,* 107: 907–938. doi: 10.1289/ehp.99107s6907.

[78] Esplugas, S., Bila, D. M., Krause, L. G. T., & Dezotti, M. (2007). Ozonation and advanced oxidation technologies to remove endocrine disrupting chemicals (EDCs) and pharmaceuticals and personal care products (PPCPs) in water effluents. *Journal of Hazardous Materials*, 149, 3: 631–642. doi: 10.1016/j.jhazmat.2007.07.073.

[79] Focazio, M. J., Kolpin, D. W., Barnes, K. K., Furlong, E. T., Meyer, M. T., Zaugg, S. D., Barber, L. B., & Thurman, M. E. (2008). "A national reconnaissance for pharmaceuticals and other organic wastewater contaminants in the United States—II) Untreated drinking water sources." *Science of the Total Environment,* 402 : 201–216. doi: 10.1016/j.scitotenv.2008.04.028.

[80] Swati, M., Rema, T., & Joseph, K. (2008). "Hazardous organic compounds in urban municipal solid waste from a developing country." *Journal of Hazardous Materials,* 160 : 213. doi: 10.1016/j.jhazmat.2008.02.111.

[81] Nowotny, N., Epp, B., vonSonntag, C., & Fahlenkamp, H. (2007). "Quantification and modeling of the elimination behavior of ecologically problematic wastewater micropollutants by adsorption on powdered and granulated activated carbon." *Environmental Science and Technology,* 41, 6: 2050–2055. http://library.dphen1.com/Documents/Papers/Nowotny-EST-2007.pdf.

[82] Snyder, S. A., Adham, S., Redding, A. M., Cannon, F. S., DeCarolis, J., Oppenheimer, J., Wert, E. C., & Yoon, Y. (2007). "Role of membranes and activated

carbon in the removal of endocrine disruptors and pharmaceuticals." *Desalination*, 202, 1–3: 156–181. https://doi.org/10.1016/j.desal.2005.12.052.

[83] Yoon, Y., Westerhoff, P., Snyder, S. A., Wert, E. C., & Yoon, J. (2007). "Removal of endocrine disrupting compounds and pharmaceuticals by nanofiltration and ultrafiltration membranes." *Desalination,* 202, 1-3: 16-23. https://doi.org/10.1016/j.desal.2005.12.033.

[84] Grassi, M., Belgiorno, V., &Lofrano, G. (2012). "Removal of emerging contaminants from water and wastewater by adsorption process." In *Emerging Compounds Removal from Wastewater,* Edited by G. Lofrano, 15-37: Springer, Dordrecht,New York, NY. https://doi.org/10.1007/978-94-007-3916-1_2.

[85] Kumar, S. (2010). "Engineering cytochrome P450 biocatalysts for biotechnology, medicine and bioremediation." *Expert Opin. Drug Metab. Toxicol.*, 6: 115–131. doi: 10.1517/17425250903431040.

[86] Clark, J.H. (2006). "Green chemistry: today (and tomorrow)." *Green Chem.*, 8: 17–21. doi: 10.1039/b516637n.

[87] Kummerer, K., & Hempel, M. (2010). "*Green and Sustainable Pharmacy.*" Edited by Heidelberg/New York: Berlin; London: Springer. https://www.worldcat.org/title/green-and-sustainable-pharmacy/oclc/663093699.

Chapter 2

A Comprehensive Insight into Historical, Structural and Mechanistic Aspects of Some Commonly Used Pharmaceutics and Their Impact on Human Health

Rajeev Singh Yadav[1,*] and Neha Agarwal[2]

[1]Department of Chemistry, Ganna Utpadak P.G. College, Baheri, Bareilly, India
[2]Department of Chemistry, Navyug Kanya Mahavidyalaya, Lucknow, India

Abstract

The concept of the term "drug" has been derived from the French word "drogue", which means a dry herb or drug. A drug is a substance that may be used or intended to be used for modifications of physiological systems for the benefit of the recipient. When human beings came into contact with natural herbs, they discovered drugs accidentally. Historically, drugs have been used in ancient times and in folklore for the treatment of various diseases in natural form. Many of these natural drugs became current drug candidates. Many medicines have found their roots in our day-to-day lives. For example, the leaves of *Ocimum sanctum* (an herb which cures common cold) can be attributed to the Hindu religious believers. Similarly, the antimalarial plant *Dichroa febrifuga* was discovered in China (2735 B.C.) by Ch'ang Shan, serendipitously. The administration of drugs is always followed by the expected reaction. There are many reactions associated with the expected reactions known as side reactions, such as mild skin rashes to serious complications like blood dyscrasia (abnormality in the blood picture), and liver damage. Some drugs also show acute and chronic toxicity and persistence in the environment. Many times, a drug possesses more than one

* Corresponding Author's Email: dr.rsyd31@gmail.com.

In: Pharmaceuticals: Boon or Bane
Editor: Neha Agarwal
ISBN: 979-8-88697-487-4

pharmacological activity. The main activity is called a 'therapeutic effect' while the rest of the actions are known as side effects of the drug. Therefore, it is very crucial to know the side effects of commonly used pharmaceuticals, their history, structural aspects, mechanism of action, and other related aspects. The objective of this chapter is to understand the beneficial and therapeutic effects of different drugs and to minimize their undesirable side-effects.

Keywords: drug design, pharmaceuticals, side effects of drugs, chronic toxicity

1. Introduction

Many researchers have proposed that organic synthesis lies at the heart of science, and many follow the rationality of Paul Ehrlich (1854-1915), the German physician and immunologist. Paul and his co-workers coined the intellectualism of medical science, such as receptors for drugs. In 1935, Domagk's discovery proposed prontosil as an antibacterial agent, which produced the basic ideas of the concept of biochemical methodologies and steric analogues. The discovery of drug design was based on modification of the structure of the 'lead' molecule. The concept of isosterism, for designing bioactive compounds, was convenient in designing a set of compounds with the same biological activities that are called bioisosterism. In 1964, Corwin Hansch discovered the concept of quantitative structure activity relationship (QSAR) that has helped medical chemists target the designing of drugs using Hammett's constants.

Compounds synthesized by chemists have played an important role in improving the quality and life of drugs for over 100 years. In the chemical industry, chemists synthesized pharmaceuticals, polymers, pesticides, detergents, and disinfectants. Researchers synthesized compounds from natural products whose structures are not certain [1-4]. For synthesizing a target molecule, a synthetic route must be designed to give a pure sample of the target molecule using a convenient and efficient procedure. A given synthetic route is a sequence of mechanistic reactions designed to convert commercially available starting materials into the desired product. In general, the best synthesis involves the conversion of the most easily available and cheapest starting material into the desired product by the least number of steps and with the highest possible yield.

An ideal drug should satisfy the administration to the ailing individual or host, and its action should be localized at the desired site of action. In medical practice, there is no drug that behaves in this manner. Normally, care has to be taken to distribute the drug anywhere in the tissues of the host body. It may act on a type of system with efficiency and safety [5]. The drug should not have any type of eco-toxicity and should have minimal side effects. The cells should not acquire tolerance or resistance to the drug after some time. In actual practice, the cells that were originally susceptible to the action of a particular drug may, after some time, acquire tolerance and resistance to that drug. Very few drugs satisfy these conditions [6]. A biologically active drug which by latentiation is converted into an inactive carrier form is called a pro-drug. A pro-drug is a pharmacologically inactive derivative of an active drug that reacts with an enzyme or non-enzyme compound that releases the active compound (drug). The concept is highly useful and is an important aspect of medicinal chemistry.

Chemistry and biochemistry have a tremendous influence on medicinal chemistry and drug research in many ways [7]. Medicinal chemistry has been aptly defined by Dr. Glen Ullyot as a field that applies the principles of chemistry and biology to the creation of knowledge, leading to the introduction of new therapeutic agents, Hence, medicinal chemistry is closely related to other branches of science and disciplines, which is indicated in Figure 1. The objective of this chapter is to improve the knowledge of medicinal chemistry so as to understand the therapeutic effects of different drugs and to minimize undesirable side-effects [8].

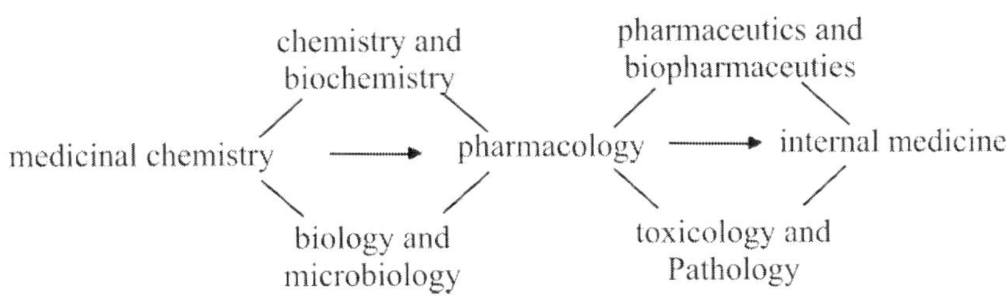

Figure 1. Relationship of medicinal chemistry to other disciplines.

2. The Evolution of Drugs

The earliest references to medicine preparations in writing come from Rigveda (India) and from Materia Medica (China) in 2500-3000 B.C. A large number of medicines have been prepared in ayurveda and their uses were explained by physicians like Charak, Sushruta, Vagbhata and others. All the drugs were of vegetable origin. A large number of drugs have been found to be listed on papyrus, and many clay tablets have been discovered in Egypt. It is evident that the 'Western medicinal system' came from Egypt and from the Kingdoms of Assyria and Babylonia [9-11]. Hippocrates (450 B.C.), the Greek physician, laid down the foundation of modern medicine. According to him, a disease is a pathological process and its treatment with drugs is not magic. The medical system introduced by Hippocrates had a scientific basis, i.e., it was based on observation, analysis, and deductions. However, up to the 19th century, it was believed that the treatment of diseases was mainly based on a combination of guesswork and experience. This resulted in quack doctors in many countries. These doctors used therapeutic preparations composed of a limited number of substances extracted from herbs and animals or from the earth in the form of minerals [12].

The era of synthetic drugs had to wait till the technique of synthetic organic chemistry became well advanced and the physiology of human organisms became well known. The beginnings of synthetic medicinal chemistry took place in Germany. Some of the earliest efforts to find synthetic drugs were concentrated on anesthetics, hypnotics, and analgesics. For example, chloral hydrate $CCl_3CH(OH)_2$ (1869), and sulfones (1888) were used as hypnotics. Phenacetin was introduced in 1889, and its discovery was based on the observation of the hydroxylation and conjugation of aniline in the animal body. It is probably the first drug to be prepared from the knowledge of biochemical transformations [13]. It was introduced as an analgesic (pain reliever). In 1899, aspirin was discovered due to some investigations aimed to reduce the nausea caused by some salicylates, which were then used as antipyretics, i.e., agents for reducing fever [14].

Table 1. Some important discoveries of drugs

S. No.	Discovery Year	Name of Drug	Category
1	1806	Morphine	Hypnotic agent
2	1899	Aspirin	Analgesic and antipyretic agent
3	1922	Insulin	Anti-Diabetic agent
4	1928	Penicillin	Antibiotics
5	1960	Chlordiazepoxide	Tranquilliser
6	1971	L-Dopa	Anti-Parkinson agent
7	1987	Artemisinin	Anti-malarial agent
8	1999	Zanamavir Oseltamivir	Anti-influenza agent
9	2001	Imatinib	Leukaemia treatment

This systematic research in pharmaceutical laboratories has led to the introduction of more and more synthetic drugs in modern times. Attempts were made to prepare the compounds with simplified structures based on the structures of natural drugs. For example, the structure of a compound having specific physiological activity is varied systematically [15]. This work sometimes led to the discovery of new drugs with a simplified structure. For example, synthetic novocaine, a local anesthetic, was discovered in this way and is used in place of naturally occurring cocaine, which shows many harmful side effects. The structures of cocaine and novacocaine are mentioned in Figure 2, and some other drugs that were discovered are mentioned in Table 1.

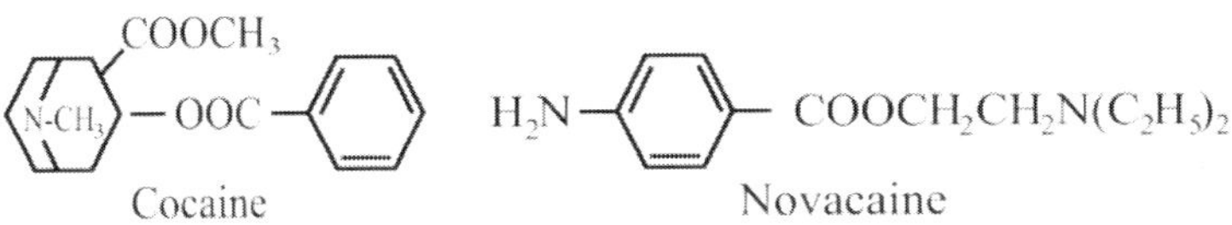

Figure 2. Structures of cocaine (left) and nova-cocaine (right).

3. Biological Transformation of Drugs

The drugs can be converted into pharmacologically active metabolites by biological transformation. This activity contributes to the pharmacological effect ascribed to the parent drug. All the metabolites are not non-toxic because many toxic side effects like tissue necrosis and carcinogenicity have been observed. The formation of water-soluble metabolites enhances drug elimination and pharmacologically inactive, non-toxic and polar compound formation. Most drugs undergo metabolic transformation in the body. The main site of metabolism is the liver [16]. Genetic differences, species differences, age, sex, nutritional and hormonal factors, pathological states, and stress influence the rate and extent of drug biotransformation and, thus, influence the duration and extent of pharmacological activity.

4. Routes, Administration and Dosage of Drugs

The speed of drug absorption in the body is determined by the route of administration. The common routes in the increasing order of rapidity of absorption in men are oral, subcutaneous, intramuscular, inhaled, and intravenous. The speed of absorption of drugs can be altered by a number of pharmacological and chemical manipulations. The injected drug is carried away from the injection site by blood circulation. During blood circulation, the drug passes through many barriers, survives and reaches the site of action (cell receptors) where the required biological response is developed. Drug administration is done by the intravenous, intra-arterial, intraspinal, and intracerebral routes, which introduce the drug directly to the body fluids and thus avoid complications in drug absorption [17]. Drugs are prepared and used in chemically pure form, and doses of drugs are calculated in appropriate weight units. Figure 3 is the schematic representation of drug pathways in the body.

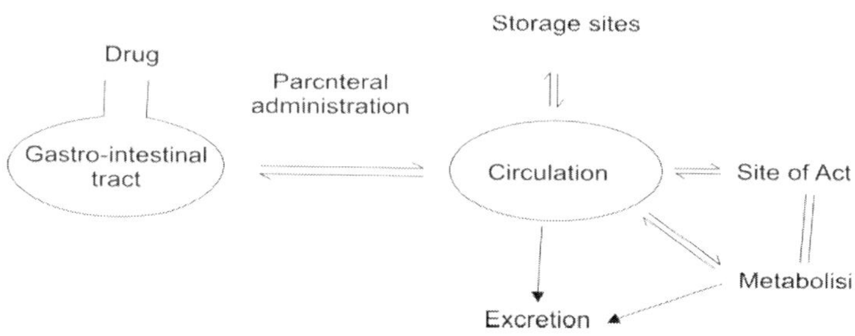

Figure 3. Schematic representation of drug administration of drugs followed by the transfer of drugs from various compartments.

5. Binding of Drugs

5.1. Drug Receptor Complex

Extensive drug binding in the body occurs in the blood. Blood contains 6.5 percent of protein, with albumin accounting for 50 percent. It is mainly involved in drug binding. Its molecular weight is 69000 and it has a net negative charge at blood pH 7.4. It can interact with anions and cations also. The drug protein binding may be due to ion-ion interactions, hydrogen bonding, hydrophobic, and Vander Waals forces. This protein-drug binding is usually a reversible reaction. The drug binding resembles salt formation. This protein binding acts as a transport system for the drug, which while bound, gets hindered in its access to the site of metabolic action and excretion [18] as in Figure 4.

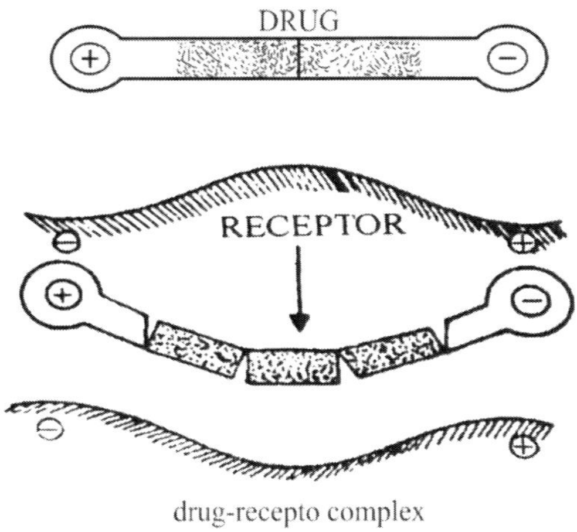

Figure 4. A representation of drug receptor complex.

(P) + (D) ↔ (PD)

Where,

(P) = Free protein concentration
(D) = Free drug concentration
(PD) = Drug-protein complex

5.2. Receptor Site

This concept can be easily understood if one considers the 'lock and key' principle for ligand-receptor interaction. Only those keys (ligands) that fulfil all types of fitting into the lock (receptor) can open the door. Some types of keys can fit into the lock but not accurately, and they cannot open the door easily. By fitting into the lock, this key prevents the original key from fitting into the lock and opening the door [19]. The concept of the key-lock receptor system is extremely important in medicinal chemistry and also underlies the mechanism of drug action. For example, to prevent the constriction of airway smooth muscle in asthmatic conditions, one can administer receptor antagonists that prevent the actions of the signalling molecules from causing muscle contraction (for example, histamine and leukotriene antagonists). Figure 5 shows the lock and key model of ligand-receptor interaction.

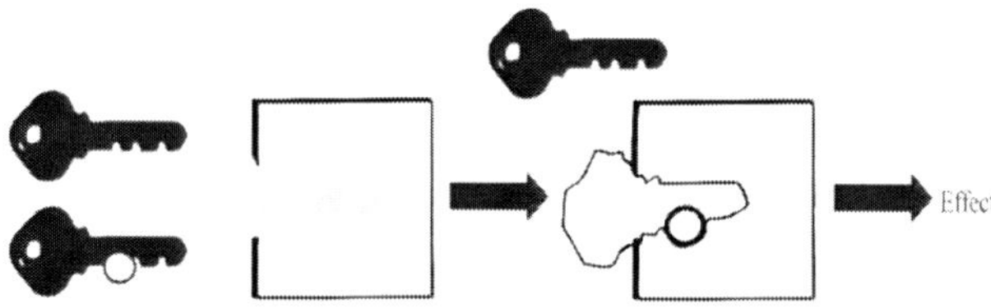

Figure 5. Ligand-receptor interaction (Lock and Key Model).

5.3. Receptor Theory of Drug Action

The receptor system is the target (usually a protein). A drug molecule binds together in order to cause a biological effect. Emil Fischer (1894) proposed that the analogy of a drug molecule is like a key that fits into a lock [20]. Ehrlich and his co-workers coined the term "receptor" in 1909 to describe the target molecule. The structure of the target can be identified and characterized by X-ray crystallography and NMR spectroscopy. For example, in 1988, the role of HIV protease was established and its amino acid sequence was determined [21].

Bond Type	Strength (kcal'mole)	Example
Covalent	40-140	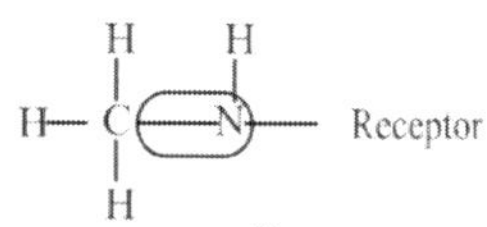
Ionic (In solution form)	5 - 10	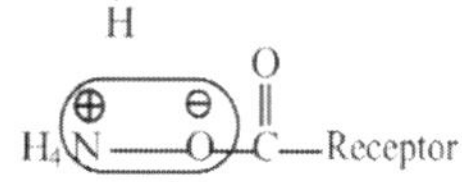
Hydrogen	1-7	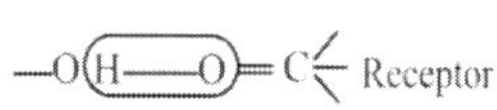
Dipole dipole	1-7	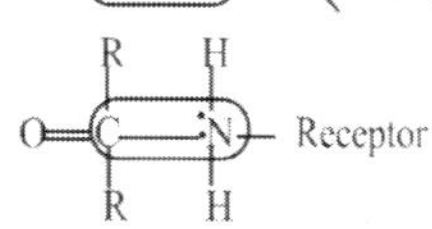
Hydrophobic	1	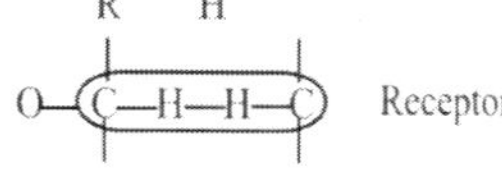

Figure 6. The binding of drugs to receptors.

The effect produced by a drug on interacting with a specific cellular component, i.e., a 'receptor', forms a drug-receptor complex. This drug receptor interaction not only provides information regarding the initial phase of drug action but also accounts for the specificity of the drug. The binding of most of the drugs to receptors does not occur through covalent bonds but by relatively weak forces, i.e., hydrophobic forces, Vander Waals forces, ionic bonds, and dipole-dipole interactions as shown in Figure 6.

5.4. Drug-Receptor Interactions

The majority of drugs show a high correlation of structure and specificity to produce pharmacological effects [19]. In most cases, the specific chemical structure is required for the receptor site and the drug structure. The binding forces of drug-receptor interactions are of the following types:

5.4.1. Covalent Interactions

By definition, the drug-receptor interaction is reversible; covalent bond formation is rather rare except in a toxic situation.

5.4.2. Ionic Interactions

Since many drugs contain acid or amine functional groups, which are ionized at physiological pH. These are strong electrostatic interactions (5-10 k.cal/mole) that are responsible for the relative orientation of the drug to its binding site [22].

5.4.3. Hydrogen Bonding Interactions (Non-Ionic or Neutral)

The drug-receptor reaction is essentially an exchange of the hydrogen bond between a drug molecule, the surrounding water, and the receptor site. The hydrogen bond strength is distance dependent and may range from 5-7 kcal/mol depending on the binding environment.

5.4.4. Vander Waals Interaction

These forces have interactions at close range and are less frequent than hydrophobic forces. These interactions are much weaker (~0.5-1 kcal/mol.) than other electrostatic interactions, and attractive forces are found over a larger surface area i.e., at the interface of ligand and binding site.

5.4.5. Lipophilic Interactions

Finally, hydrophobic bonds are formed between non-polar hydrocarbon groups on the drug and those in the receptor site.

6. Representative Drugs

6.1. Antipyretics

The drugs that help to reduce fever to a normal body temperature are called antipyretic drugs. These antipyretic agents also have mild analgesic activity [23]. Antipyretics (such as aspirin) have been widely used since the late 19th century, but their mechanism of action by which they relieve fever has only been characterized in the last few decades. It is now clear that most antipyretics work by inhibiting the enzyme cyclooxygenase and reducing the levels of PGE_2 within the hypothalamus.

6.2. Analgesics

Analgesics are agents that relieve pain by elevating the pain threshold without disturbing consciousness or altering other sensory modalities. Analgesics are classified into the following two groups:

6.2.1. Narcotics

They produce analgesia and sleep and, in high doses, cause unconsciousness. They are mainly the opium alkaloids or their synthetic and semi-synthetic derivatives, e.g., Morphine, Codeine, Pethidine, etc.

6.2.2. Non-Narcotics

These are the drugs that are not so potent and do not cause addiction. They usually produce an antipyretic effect as well. e.g., Aspirin, Analgin.

Some important examples of these analgesic-antipyretic drugs are described below.

a. Aspirin

It is also known as Acetyl Salicylic Acid. It was introduced into medicine by Dresser in 1899. It was first prepared by Kolbe in 1774 by refluxing salicylic acid with either a mixture of acetic anhydride and acetic in the presence of conc. H_2SO_4 or with acetic anhydride in the presence of pyridine [24]. Aspirin is used as an antipyretic, analgesic, and antirheumatic, usually in powder, capsule, suppository, or tablet form. There is some anaesthetic action when applied orally. Aspirin causes several side effects. It may cause minute ulcers in the stomach and may cause bleeding. Some people are allergic to aspirin. The usual allergic reaction is rashes on skin, lowering of blood pressure, bronchial spasm, profuse sweating, intense thirst, nausea and vomiting [25]. Aspirin is synthesized from salicylic acid as shown in Figure 7.

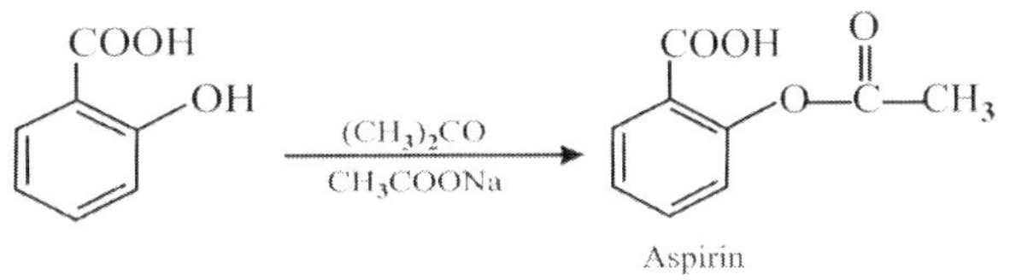

Figure 7. Formation of aspirin.

Aspirin has many side effects which may be summarized as –

- At analgesic doses (0.3-1.5g/day), aspirin produces nausea, vomiting, epigastric distress, and increased occult blood loss in stools.
- The most important adverse effects of aspirin are gastric mucosal damage and peptic ulceration.
- Aspirin produces some adverse reactions, including rashes, urticaria, rhinorrhea, asthma, dizziness, vertigo, excitement, and mental confusion.
- In children, aspirin produces acute salicylate poisoning.

b. Paracetamol

It is used as an analgesic and an antipyretic drug. The other name for paracetamol is acetaminophen. It is used to treat mild to moderate fevers and pain. Paracetamol is available in pure form as numerous trade-name preparations for oral use. It is also found in combination with many other drugs. It is preferred over aspirin, particularly in patients with coagulation disorders, individuals with a history of peptic ulcers or who cannot tolerate aspirin, and in children. It is synthesized by the acetylation of p-nitrophenol as shown in Figure 8.

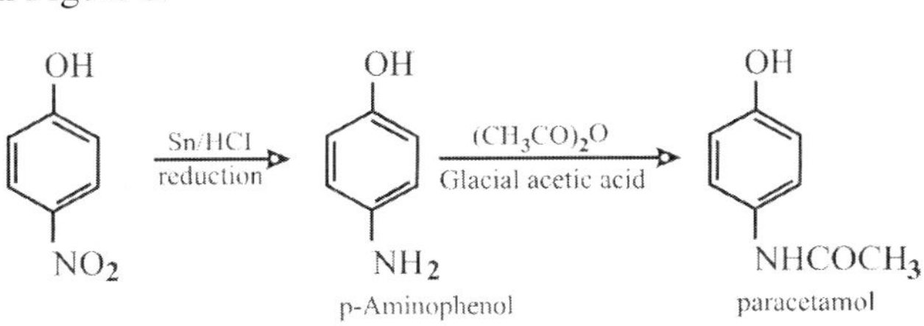

Figure 8. Showing formation of paracetamol.

It has many side effects, such as nausea, rashes, and leucopenia, but it is rare. Acute paracetamol poisoning occurs specifically in small children who have low hepatic glucuronide conjugating ability. Overuse of paracetamol by a mother during pregnancy may cause behavioral disorders in children, though much research is required to come to a concrete conclusion.

6.3. Antibiotics

Antibiotics are those microbial metabolites which can inhibit the growth of microorganisms. 2,500 years ago, true antibiotics were recognized in Chinese folk medicine, when the beneficial effects of medicine made from mouldy bean curd were reported. In the 19th century, bacteriology became a science when these substances were studied systematically. Pasteur and Joubert (1877) discovered that introducing "common bacteria" into a pure culture of *Anthrax bacilli* caused the bacilli to die, and that injecting dead anthrax bacilli into a laboratory animal along with common bacteria was harmless [26]. The main focus on antibiotics was clearly brought after A. Flemins's discovery of penicillin in 1929. In simple words, "an antibiotic is a chemical that is produced by a living organism in a comparatively high dilution and inhibits the growth or reproduction of some other microorganisms." Some important antibiotics are described as follows.

a. Chloramphenicol

One of the most commonly used antibiotics is chloramphenicol, or chloromycetin. It was first isolated from *Streptomyces venezuelae*. The structure of chloramphenicol is shown in Figure 9. It is the first antibiotic produced synthetically on a commercial scale.

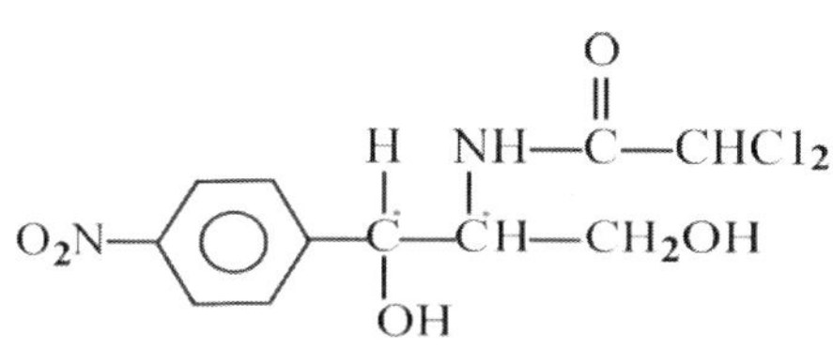

Figure 9. Structure of chloramphenicol.

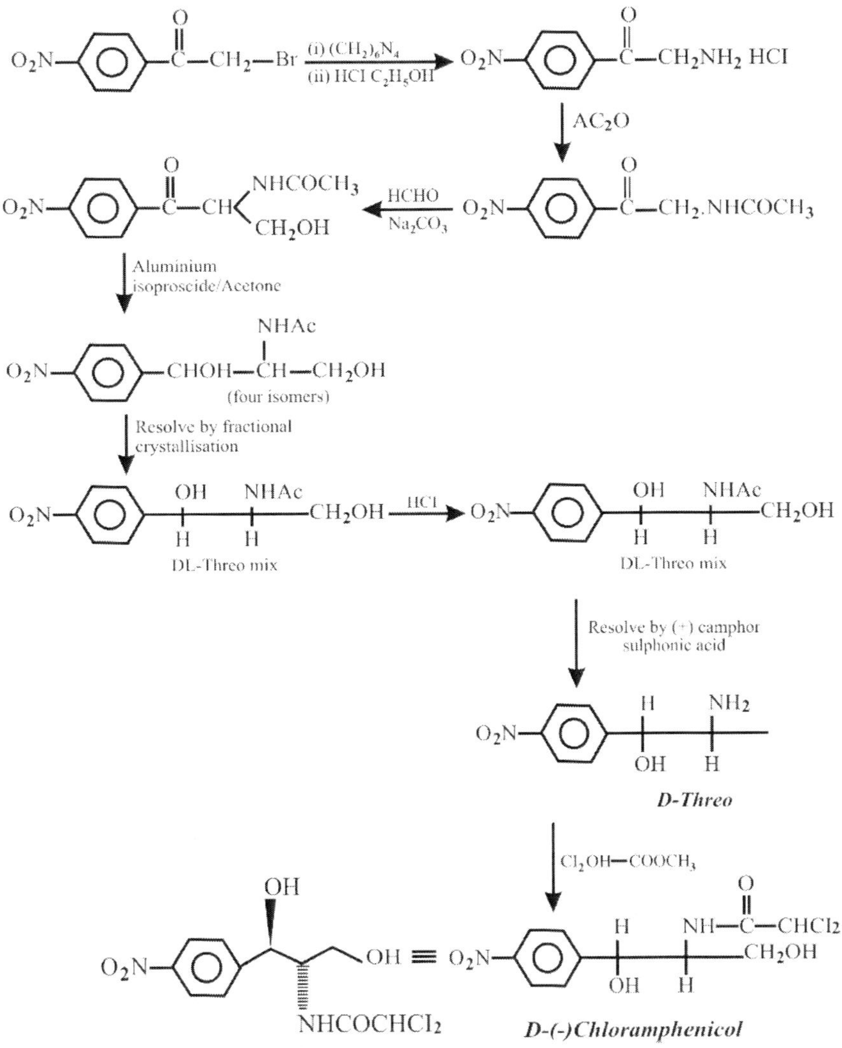

Figure 10. Commercial preparation of chloramphenicol.

- *Mode of Action*: Chloramphenicol inhibits bacterial protein synthesis by interfering with "transfer" of the elongating peptide chain to the newly attached aminoacyl t-RNA at the ribosome m-RNA complex. It attaches to the 50 S ribosome and hence hinders the access of aminoacyl t-RNA to the acceptor site for amino acid incorporation, most probably by acting as a peptide analogue. It prevents the formation of peptide bonds [27]. At high doses, chloramphenicol may inhibit the mammalian mitochondrial protein synthesis. This type of drug is rapidly and fully absorbed after its oral administration.

- *Uses and the adverse effects*: It is a drug of choice for typhoid fever and is also used for intraocular infections caused by *H. influenzae meningitis*, anaerobic infections caused by *B. fragilis,* and others. It is highly effective topically, e.g., for external ear infections. This type of drug has been extensively used in urinary tract infections as well. Side effects of chloramphenicol include bone marrow depression, rashes, fever, angioedema, atrophic glossitis, nausea, vomiting, diarrhea, pain, and aplastic anaemia [28]. Commercially it is prepared as per the mechanism described in Figure 10.

b. Streptomycin

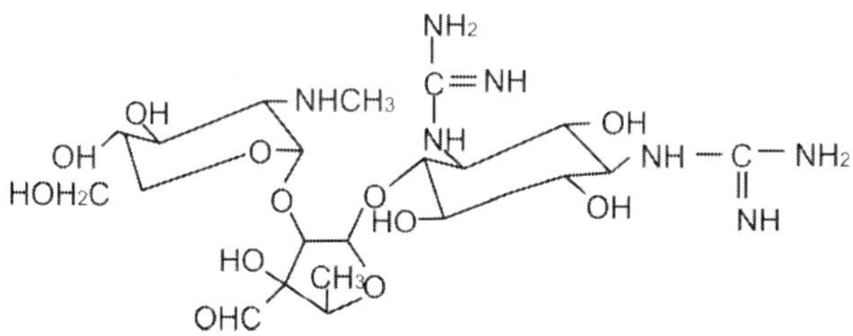

Figure 11. Structure of antibiotic streptomycin.

It is an antibiotic, first isolated from *Streptomyces grises* by Schatz, Bugie, and Waksman in 1944. It is very effective in the treatment of tuberculosis, meningitis, and pneumonia. Streptomycin is a colorless solid, soluble in water, and is laevorotatory in nature [29]. It is stable at or below 28 degrees celsius and between pH 3 and 7. It is not obtained as a crystalline compound. Streptomycin is an amino base and forms salts that are highly soluble in water; the sulfate and the calcium chloride double salts are white amorphous powder. The molecule contains two strongly basic guanidino groups and weakly basic methylamino groups. Figure 11 shows the structure of streptomycin.

- *Clinical uses of streptomycin*: It is a bactericidal drug which is used in the treatment of pulmonary tuberculosis, plague, and tularaemia. It is also useful in infections produced by *E. coli., P. pyocyaneus, H. influenzae, Proteus vulgaris, and Brucella abortus.* Aside from tuberculosis, it is used with penicillin to treat subacute bacterial endocarditis caused by *Streptococcus faecalis*, resistant staphylococci, and post-operative infections [30]. It is a useful drug for treating pneumonia and urinary tract infection. Streptomycin is rapidly absorbed after intramuscular injections. Oral administration of streptomycin is of no value in the treatment of systemic infection [31].

- *Mechanism of Action of Streptomycin*: While the mechanism of action of streptomycin against bacteria remains obscure, a number of investigations have demonstrated that this antibiotic interferes with normal cellular division. Streptomycin intervenes in ribonucleic acid metabolism by blocking a reaction preceding the formation of mononucleotides [32]. Streptomycin exerts its effects on tubercle bacilli when they are in a growing rather than a resting phase [33]. Some medical chemists reported that the process of division with the rapid synthesis of bacteria cells is blocked by the drug. Among many antibiotics isolated from the genus Streptomyces, there are a number of compounds, such as Kanamycin, Neomycin, Paromomycin, and Gentamicin, which are used to isolate antibiotics [34]. The structure of neomycin is given in Figure 12.

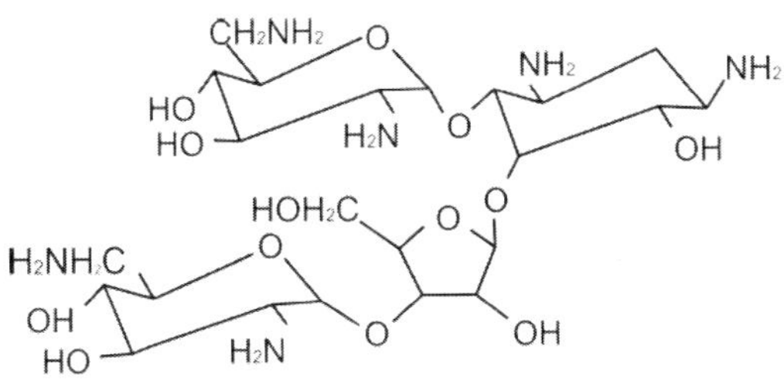

Figure 12. Structure of antibiotic neomycin C.

6.4. Antineoplastics

Antineoplastic agents (anticancer drugs) represent a diverse class of medications. They generally have limited but important uses and often have significant hepatotoxicity. These are categorized anti-metabolites, alkylating agents, hormones and antagonists, natural products, etc. They are used for the treatment of cancer. Cancer is a form of abnormal development that transforms normal cells into cancerous cells. Cancer is a tumour, which means an unusual amount of growth or enlargement of a tissue due to unlimited and uncontrolled repeated divisions of some cells [35]. This kind of development usually occurs after the age of 35-40 years, but it may also occur at a younger age. Generally, the cancerous tumour passes through a latent period of somewhat slow growth when the symptoms are not very obvious. Its growth occurs rapidly after the age of 50 and also causes the death of individuals, but not all tumours are cancerous. Two general types of tumours are recognized [36]. A benign (non-malignant) tumour grows slowly but it becomes quite large. It is restricted to the area of its origin and does not spread out to other areas of the body. It does not give rise to cancer. Most tumours are of this type. Another type of cancerous (malignant) tumour also begins as a small growth at first, for example, a pigmented mole somewhere on the body or a small swelling on the breast. It can grow slowly at first and more rapidly after some time. The tumour ultimately starts spreading to the neighbouring tissues, like the roots of a tree. The last stage comes when the cells of the tumour break away from it and migrate through the bloodstream or lymph to the other parts of the body. There they accumulate and form secondary tumorous growths [37]. This type of stage is also called meta-stasis, which is fatal and causes death sooner or later. This cancer may therefore be defined as unorganized growth in which the controlling and regulating mechanisms have disappeared or have become ineffective.

6.4.1. Cancer-Causing Agents

It is still not clear how cancer is caused. Nowadays, some factors are known to be cancer producing or carcinogenic. They are listed as follows:

- If a person smokes heavily, this may cause constant irritation to the tissues of the lungs, which may lead to cancer. For this reason, heavy smoking is found to be injurious to health.
- Constant or chronic physical irritation of any tissue in the body is considered to be one of the causes of cancer. Kashmiris have the habit of keeping a pot of burning coal (kangiri) against their abdomens to protect them against the cold during winter. As the heat is constantly irritating the abdominal tissues, this may be one of the causes of the greater frequency of abdominal skin cancer among the Kashmiris.
- Chewing of pan or tobacco causes continuous irritation to the buccal epithelium. This irritation causes mouth cancer. For this valid reason, the excessive chewing of betel, especially keeping the tobacco in the mouth for a longer period, is quite harmful for health [38]. These carcinogenic substances include nicotine, caffeine, products of combustion of coal and oil, several polycyclic hydrocarbons, some sex hormones, and steroids if given or produced in large amounts.

- X-rays, ultra-violet rays, and other ionizing radiation.
- Viruses

Cancer does not occur in humans alone; it can also develop in many other mammals and lower vertebrates, and even in insects. Cancerous growths also occur in plants, but in them they are caused by bacterial infections which have not been found to cause cancer in animals [39]. Two types of hypotheses have been suggested for how a normal cell changes into a cancerous cell. According to one hypothesis, cancer may be essentially due to alterations in one or more chromosomes or gene mutations in the nuclei of the somatic cells. According to the other hypothesis, genetic change may not be an essential factor. Instead, some change occurs in the cytoplasm, resulting in the loss of control of the nuclear and cytoplasmic division [40].

6.4.2. Cancer Treatments

Cancer is a very dreadful disease. It is one of the biggest killers, next only to cardiovascular diseases. A scientific effort has been employed to trace the nature, cause, and cure of this type of disease. The following methods are generally used in the treatment of cancer;

a. *Surgery*: Surgical methods are generally used for the removal of cancerous cells. However, this type of method is not reliable because, if any cancerous cells are dislodged during operations, they may form secondary growth of the cells. Hence, this method is not a reliable method for the treatment of cancer.
b. *Radiation Therapy*: In this method, the cancerous growth is destroyed by lethal radiation like x-rays, gamma-rays, etc.
c. *Chemotherapy*: As far as chemotherapy is concerned, at present, no available compound could cure any form of cancer. However, many compounds have been found that could cure any form of cancer. However, many compounds have been found which are able to provide relief from pain, increase survival time, prevent the spread of disease and, in some cases, even bring about the temporary disappearance of tumours [41]. The chemotherapeutic agents that are also used in the treatment of cancer may be described as follows:
d. *Cytotoxic Agents*: These are characterized by containing at least one and usually two or more reactive groups in the molecule and are capable of introducing alkyl groups into the molecule that perform

important cellular functions. They are often known as polyfunctional alkylating agents [42]. Effective cytotoxic agents are injurious to rapidly proliferating cells, which are present in many tumours. Some common cytotoxic agents are described as follows:

e. *Mustards*: During World War I (1914-1919), mustard gas was used as a poisonous gas. It was observed that it had a damaging effect on the lymphatic tissues and bone marrow. Therefore, it was found to be of some value in the treatment of certain types of skin cancer. However, due to its toxicity and low solubility in water and the fact that it produces blisters; its clinical use was prevented [43]. This type of preparation of nitrogen analogues of mustard gas during World War II was found to be effective as antineoplastic agents. For example, mechlorethamine hydrochloride (mustargen, musteric hydrochloride). It is a hydrochloride of 2,2'-dichloro-N-methyl diethylamine and is also known as nitrogen mustard. Different mustard compounds are shown in Figure 13. It is a white hygroscopic powder that is stable up to 44°C. It is very rapidly converted into its active form when brought into contact with water, and therefore its solution is prepared just before its use [44]. In aqueous solution, it undergoes rapid intramolecular rearrangement with loss of Cl^- ion to form a cyclic onium cation which may release alkyl groups as shown in Figure 14.

Mechlorethamine hydrochloride is administered in doses of 0.1 mg/kg once daily for four consecutive weeks. The course of treatment may be repeated at a 4-week interval. Generally, the maximum number of courses is 5. Contact with the skin and mucous membranes must be avoided [45]. Mechlorethamine hydrochloride is used in the treatment of Hodgkin's disease (enlargement of lymphatic glands) and lymphosarcoma (a lymphatic cancer). Beneficial results have also been reported in the cases of carcinomas of the bronchi, breasts, and ovaries. The biological effect of mechlorethamine hydrochloride may be considered specifically cytotoxic. Cellular enzyme systems are inhibited and cell death ensues [46]. Mechlorethamine hydrochloride has side effects, and therefore its usage is restricted to the treatment of highly specific types of cancer. Other similar types of compounds are shown in Figure 13.

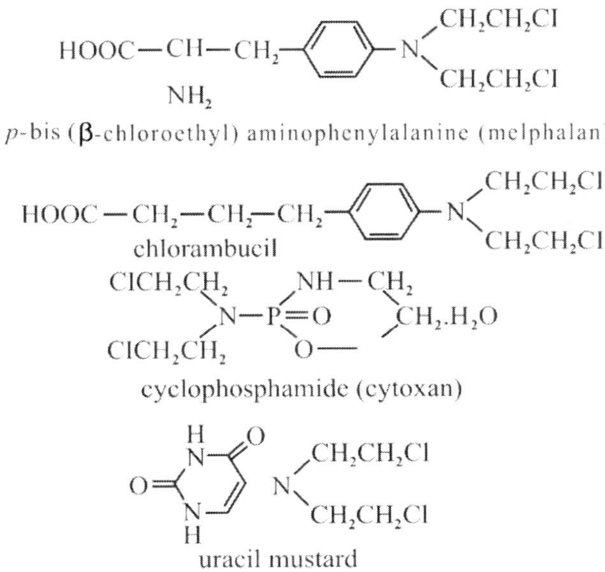

Figure 13. Different mustard compounds.

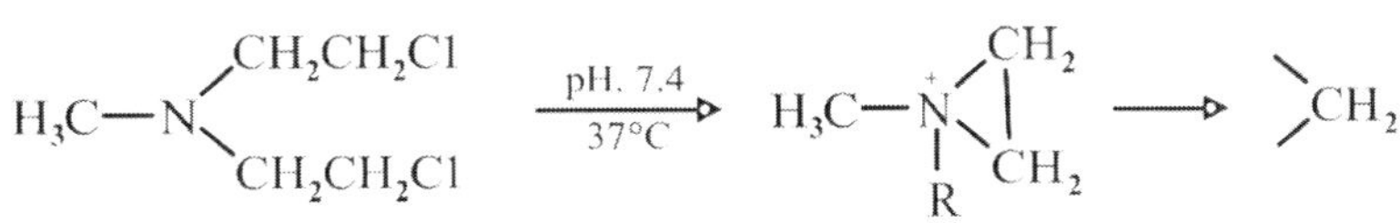

Figure 14. Releasing of alkyl group.

6.5. Antiviral Agents

These antiviral agents are the drugs that are approved by the Food and Drug Administration (FDA) for the treatment and disease control of viral infections. Some antiviral agents mainly target stages in the inflected life cycle. The target stages in the inflected life-cycle are viral infected attachment to host cell, uncoating, viral mRNA synthesis, mRNA translation and viral RNA and DNA replication in body fluids. Half of the antiviral agents are available for the treatment of Human Immunodeficiency Virus (HIV) infections. The others are used mainly for the management of hypersensitivity, hepatitis B virus (HBV), hepatitis C virus (HCV) and respiratory viruses [47]. Antiviral agents can be used for prophylaxis, suppression, pre-impactive therapy, or treatment of overt disease. In addition, host phenotypic behaviors towards antiviral drugs because of either genomic or epigenetic factors could limit the efficacy of an antiviral agent in an individual [48]. These types of drugs may kill or prevent the growth of viruses. Oseltamivir (Tamiflu) and Zanamivir (Relenza) are two antiviral drugs that may be medicinally used against the flu virus. The antiviral drugs may cause symptoms and can shorten the time you are sick with viral infections like the flu and ebola. They can get your body rid of these viruses and viral infections like HIV, hepatitis, and herpes, which are chronic [49]. Viruses are obligate parasites that need the active participation of the metabolic processes of the invaded cell. Antiviral chemotherapy is confronted with two major obstacles, which explain its slow progress and scanty achievements:

- due to the intimate relationship that exists between multiplying viruses and mammalian cells; most antiviral agents lack selectivity, being equally toxic to both virus and host.
- many viral diseases are diagnosed too late for effective treatment; usually the first symptoms of viral diseases appear at the final stage of viral multiplication.

Despite the great efforts towards discovering useful antivirals, only a few are presently available. None has broad-spectrum antiviral activity, although drugs of this sort are badly needed. Therefore, antiviral chemotherapy was at the same stage as antibacterial chemotherapy before the introduction of sulfa drugs. For prophylaxis of most serious viral diseases, the only available and dependable resources are vaccines. For the treatment of others, there are some recommended preparations. Several antiviral medications, such as idoxuridine, methisazone, Amantadine Hydrochloride, cytarabine, may be discussed in this section [50].

6.5.1. Idoxuridine

It is 5-iodo-2'-deoxyuridine. It forms odourless crystals of white crystalline powder and has to be stored in well-closed, light-resistant containers. It has a slight solubility in water. Its aqueous solutions have been found to be most stable at pH 2 to 6. Its aqueous solutions are freshly prepared and stored in the refrigerator. It finds use in the treatment of superficial herpes simplex keratitis (dendritic keratitis) of recent origin. Idoxuridine is applied as an eye drop or eye ointment [51], which is shown in Figure 15.

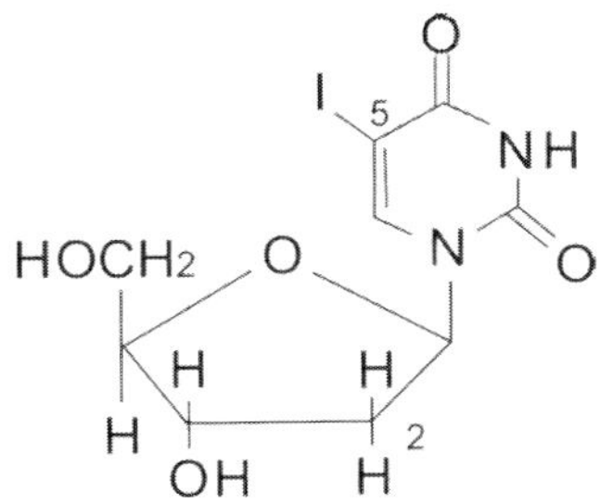

Figure 15. Structure of idoxuridine compound.

6.5.2. Methisazone (Methisazone)

Chemically, it is 1-methylindoline-2, 3-dione3-thiosemi-carbazone. It is an orange-yellow powder which is particularly insoluble in water. Methisazone should be protected from light. It has been reported to be active against the poxviruses and is being used in the prophylaxis of smallpox and alas trims (variola minor), although smallpox vaccination is regarded to provide better long-term protection. It may be used for mass administration in conjunction with vaccination in the event of an epidemic of small-pox ever occurring again. It is also given to prevent complications following vaccination and is used in the treatment of eczema vaccination and vaccinia gangrenosa [52]. It is given orally and its structure is shown below in Figure 16.

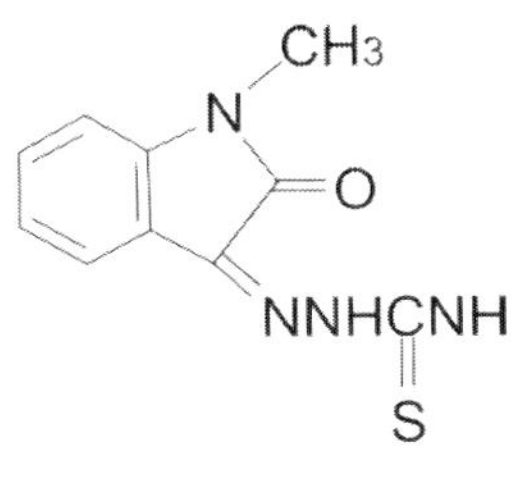

Figure 16. Structure of the methisazone compound.

6.5.3. Amantadine Hydrochloride

It is an antiviral agent which is used prophylactically against infection with the influenza A 2 viruses. It is not effective in preventing infection with other types of influenza virus. It has been tried in the treatment of herpes zoster [53]. Amantadine may increase dopaminergic activity and finds use in the treatment of parkinsonism, usually in conjunction with other therapies. Amantadine hydrochloride is given to patients orally.

6.5.4. Cytarabine

It is a pyrimidine nucleoside analogue which finds use as an antineoplastic agent, mainly in the treatment of leukemia. It has been used as an antiviral drug. It is given by intravenous injection in the treatment of herpes infection. It is given intrathecally for encephalitis [54].

6.6. HIV-AIDS Related Drugs

These fall under the category of antiviral agents, more specifically antiretroviral agents. HIV is also called a retrovirus. A retrovirus is a virus composed of RNA rather than DNA. A retrovirus has an enzyme, reverse transcriptase, which transcribes RNA into DNA inside the host cell. HIV infection in its later stages leads to the development of AIDS (Acquired Immune Deficiency Syndrome), which is characterized by severe reduction in CD4 + T cells which is life threatening [55]. One of the important antiviral drugs is zidovudine, which is also known as azidothymidine.

6.6.1. Zidovudine

It is the first antiviral drug and is used to prevent HIV-AIDS. The structure is shown in Figure 17.

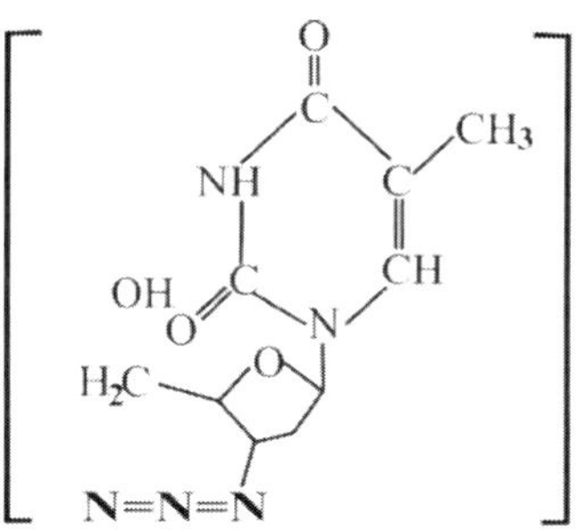

Figure 17. Structure of zudovudine (azidothymidine).

It is white or off-white in color and crystalline in nature, soluble in water and ethanol [56]. It is used to prevent HIV in women to prevent the HIV virus from passing to the baby. Zidovudine is an antiretroviral medicine used to prevent and treat HIV-AIDS. Its use is recommended in combination with other antiretrovirals.

Mode of Action

Zidovudine is the analogue of thymidine. It works by inhibiting HIV's reverse transcriptase, selectively. Reverse transcriptase is the enzyme which is used by the virus to make DNA copies from RNA. Reverse transcription is required for the production of HIV's double-stranded DNA, which is integrated into the genetic material of the infected cell [57, 58]. The protein reverse transcriptase is required for the reproduction of the HIV virus. High doses of azidothymidine triphosphate form may also inhibit DNA polymerase, which is used by cells to undergo cell division. But regardless of dosage, azidothymidine has very high affinity for HIV's reverse transcriptase [59]. The drug blocks this protein and hence helps to slow down HIV disease, as shown in Figure 18.

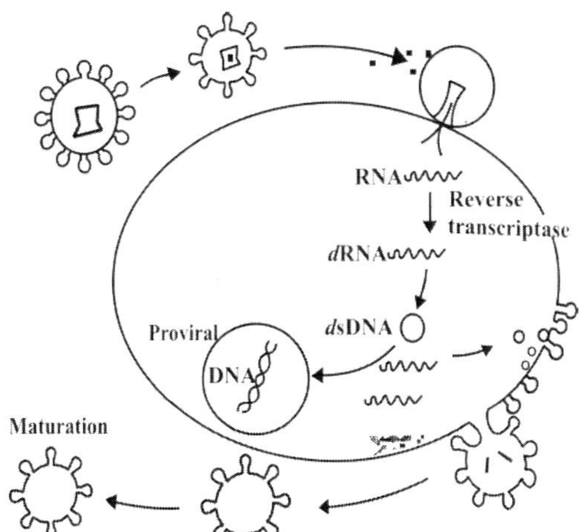

Figure 18. Virus in the attacking mode.

Side-Effects of Zidovudine

It is quite effective to prevent HIV, but it may cause many side-effects. These symptoms may be mild to severe, such as stomach aches or cramps, heartburn, diarrhoea (especially in children), constipation, headache, difficulty falling asleep or staying asleep.

Uses of Zidovudine

It is used along with many other medicines to treat HIV infection. Zidovudine is also given to HIV-positive pregnant women to reduce the chance of passing the infection to the baby. Zidovudine may be used to prevent mother-to-child spread during birth or after a needle stick or other potential exposure.

Conclusion

It is concluded that the usage of pharmaceuticals is to cure patients and relieve them from several diseases. Our traditional medicine and modern medicine are the two types of medicinal treatments which are followed by the global population. As discussed, almost all types of medicine have undesirable side effects. The medicines should be taken in a rational manner and in the correct quantity with a proper prescription so that the right medicine can be used according to the type of disease in order to avoid drug abuse side effects. The main target of all types of medicine is to help patients live long, pain-free, and healthy lives. The entire medical field and researchers can adopt the principles of evidence-based medicine so that it may be possible to minimize the side effects and also maximize the benefits of scientific research for patient care. Moreover, further development should be done in the entire medical field to improve the treatment techniques and also scientific medical research should be done for medication and treatment of incurable diseases. In the future, the medical field may achieve a further standard in the treatment method for the well-being of our everyday lives.

References

[1] History Philosophy and Theory of Life Sciences. (2016). in *"Early Modern Medicine and Natural Philosophy."* Edited by Benjamin Goldberg, Evan R. Ragland & Peter Distelzweig. 2016. 14: 1-15. Kindle Edition, Springer Nature. https://www.amazon.in/Medicine-Natural-Philosophy-History-Sciences-ebook/dp/B0199NGGRY.

[2] Halliwell, B. & Gutteridge, J.M.C., Eds. (2015). *Free Radicals in Biology and Medicine*. University Press Scholarship online. DOI:10.1093/acprof:oso/9780198717478.001.0001.

[3] Williams, S.J., & Calnan, M., Eds. (2013). *Modern Medicine, Lay Perspectives and Experiences*. Routledge Talyor and Francis Group: London. Https://doi.org/10.4324/9781315072395.

[4] Bentley, G., R. (2016). "Applying Evolutionary Thinking in Medicine: An Introduction." In *Advances in the Evolutionary Analysis of Human Behavior*, edited by Alvergne, A., Jenkinson, C., Faurie, C. 1-16: Springer, Cham. DOI: 10.1007/978-3-319-29716-3_1.

[5] Collins, & Varmus, H. (2015). "A New Initiative on Precision Medicine." *The New England Journal of Medicine*, 372: 793-795. DOI: 10.1056/NEJMp1500523.

[6] Ackerknecht, E.H. (2016). *A Short History of Medicine*. Johns Hopkins University press, Maryland. https://www.worldcat.org/title/short-history-of-medicine/oclc/941071423?referer=di&ht=edition.

[7] Perlman, R.L. (2013). "Evolution and Medicine." *Perspectives in Biology and Medicine, 56*, 2: 167-183. DOI: 10.1353/pbm.2013.0012.

[8] Perlman, R.L. (2013). *Evolution and Medicine*: Oxford University Press, UK. DOI: 10.1093/acprof:oso/9780199661718.003.0001.

[9] Stanley, J.O. (2015). "Presidential Inauguration Speech: The Evolution of Medicine." *Missouri Medicine,* 112, 3: 228-239. https://www.ncbi.nlm.nih.gov/pmc/articles/PMC6170129/.

[10] Sullivan, E. (2011). "Culture, Literature, and the History of Medicine." *Medical History,* 55, 4: 541–548. https://www.ncbi.nlm.nih.gov/pmc/articles/PMC.3199652/.

[11] Chaplin, S. (2016). "Why Creating a Digital Library for the History of Medicine is Harder than You'd Think." *Medical History,* 60, 1: 126–129. DOI:10.1017/mdh.2015.83.

[12] Nesse, R.M., & Stearns, S.C. (2008). "The great opportunity: Evolutionary applications to medicine and public health." *Evolutionary applications,* 1, 1: 28-48. doi: 10.1111/j.1752-4571.2007.00006.x.

[13] Krishna, S., Bustamante, L., Haynes, R.K., & Staines, H.M. (2008). "Artemisinins: their growing importance in medicine." *Trends in Pharmacological Sciences*, 29, 10: 520–527. DOI: 10.1016/j.tips.2008.07.004.

[14] Erden, A. (2015). "Focus: Personalized Medicine." *Yale Journal of Biology and Medicine,* 88, 4: 349. https://www.ncbi.nlm.nih.gov/pmc/articles/PMC4654182/.

[15] Kushner, R.F., & Mechanick, J.I. (2016). "The Importance of Healthy Living and Defining Lifestyle Medicine." In *Lifestyle Medicine*. 9-15. DOI:10.1007/978-3-319-24687-1_2.

[16] Armstrong, D.L. (2015). "An Outline of Sociology as Applied to Medicine." *Wright London*: 1-136. Elsevier Science. https://www.perlego.com/book/1899194/an-outline-of-sociology-as-applied-to-medicine-pdf.

[17] Buriani, A., Garcia-Bermejo, M.L., Bosisio, E., Xu, Q., Li, H., Dong, X., Simmonds, M.S.J., Carrara, M., Tejedor, N., Lucio-Kazana, J., & Hylands, P.J. (2012). "Omic Techniques in Systems Biology Approaches to Traditional Chinese Medicine Research: Present and Future." *Journal of Ethnopharmacology,* 140, 3: 535–544. Toxicity of drugs- book chapter-New.docx.

[18] Micozzi, M.S. (2015). *Fundamentals of Complementary and Alternative Medicine*: Elsevier Saunders, UK. https://evolve.elsevier.com/cs/product/9780702062131?role=student.

[19] Gawde, S.R., Shetty, Y.C., & Pawar, D.B. (2013). "Knowledge, Attitude, and Practices toward Ayurvedic Medicine Use among Allopathic Resident Doctors: A Cross-Sectional Study at a Tertiary Care Hospital in India." *Perspectives in Clinical Research,* 4, 3: 175–180. DOI: 10.4103/2229-3485.115380.

[20] Young, A. (1982). "The Anthropologies of Illness and Sickness." *Annual Review of Anthropology,* 11: 257-285. https://doi.org/10.1146/annurev.an.11.100182.001353.

[21] Rossi, E., Picchi, M., Panozzo, M., Stefano, M.D., & Baccetti, S. (2015). "Integration of Homeopathy and Complementary Medicine in the Public Health System in Italy: National Regulation and Regional Experiences." *Journal of Medicine and the Person,* 13, 1: 45-54. DOI:10.1007/s12682-014-0187-0.

[22] Whelan, G.J. (2014). "Homeopathy in the Treatment of Allergic Diseases." In *Encyclopedia of Medical Immunology*, Edited by Mackay I.R., Rose N.R., Ledford D.K., Lockey R.F: Springer, New York, NY. https://doi.org/10.1007/978-1-4614-9194-1_251.

[23] Sagi, M. (2016). "The New Homeopathy: A New Paradigm in Information Medicine, World Futures." *The Journal of New Paradigm Research,* 72, 1-2: 53-68. https://doi.org/10.1080/02604027.2016.1143292.

[24] Sharma, C. (2015). "Antibiotic Resistant Organism: An Emerging Public Health Problem and Role of Ayurveda (An Overview)." *International Journal of Ayurveda and Pharma Research,* 2, 1. https://www.semanticscholar.org/paper/ANTIBIOTIC-RESISTANT-ORGANISM%3A-AN-EMERGING-PUBLIC-Sharma-Singh/6b2c157704d686bd11cc8aa7f60c8c21d74eff48#citing-papers.

[25] Abitha, J. (2016). "Role of Antioxidants, Anti-Carcinogens and Anti-Mutagens." *DJ' Journal of Engineering Chemistry and Fuel,* 1, 3: 8-16. DOI:10.18831/djchem.org/2016031002.

[26] Pubramani, S., Thing, Gan Siaw, Sokkalingam, Dhanaraj Arumugam. (2014) *Polyherbal Formulation: Concept of Ayurveda,* 8, 16: 73-80. DOI: 10.4103/0973-7847.134229.

[27] Jaiswal, Y.S., & Williams, L.L. (2016). "A Glimpse of Ayurveda – The Forgotten History and Principles of Indian Traditional Medicine." *Journal of Traditional and Complementary Medicine,* 7, 1: 50-53. DOI: 10.1016/j.jtcme.2016.02.002.

[28] Walter, T.M., Merish, S., & Tamizhamuthu, M. (2014). "Review of Alternanthera Cassilis with Reference to Traditional Siddha Medicine." *International Journal of Pharmacognosy and Phytochemical Research,* 6, 2: 249-254. https://ijppr.openresearchjournals.com/index.php/ijppr/article/view/409.

[29] Tangwa, G.B. (2015). "Traditional Medicine." In *Encyclopedia of Global Bioethics* edited by Ten Have H: Springer Cham. DOI: https://doi.org/10.1007/978-3-319-05544-2_421-1.

[30] Raja-Rina-Raja, I., Abd Ghani, M.K., & Abdullah, N. (2015). "An Analysis of Application of Health Informatics in Traditional Medicine: A Review of four Traditional Medicine Systems." *International Journal of Medical Informatics,* 84, 11: 988-996. DOI: 10.1016/j.ijmedinf.2015.05.007.

[31] Yuan, H., Ma, Q., Ye, L., & Piao, G. (2016). "The Traditional Medicine and Modern Medicine from Natural Products." *Molecules,* 21: 559. https://doi.org/10.3390/molecules21050559.

[32] Houshmand, S., Gholami, S., Salavati, A., & Alavi, A. (2014). "The Role of Positron Emission Tomography- Computed Tomography/Magnetic Resonance Imaging in Modern Medicine." *Frontiers in Biomedical Technologies,* 1, 3. https://fbt.tums.ac.ir/index.php/fbt/article/view/6/0.

[33] More, B. (2016). "Overview of Medicine: Its importance and impact." *DJ International Journal of Medical Research,* 1, 1: 1-8. https://dx.doi.org/10.18831/djmed.org/2016011001.

[34] Fauzi, F.M., Koutsoukas, A., Lowe, R., Joshi, K., Fan, T.P., Glen, R.C., & Bender, A. 2013. "Linking Ayurveda and Western Medicine by Integrative Analysis." *Journal of Ayurveda and Integrative Medicine,* 4, 2: 117–119. DOI:10.4103/0975-9476.113882.

[35] Kelner, M., & Wellman, B. (2014). *Complementary and Alternative Medicine: Challenge and Change.* Routledge: Taylor and Francis Group. https://www.routledge.com/Complementary-and-Alternative-Medicine-Challenge-and-Change/Kelner-Wellman/p/book/9789058230997.

[36] Nicolaou, K.C. (2013). "The Emergence of the Structure of the Molecules and the Art of its Synthesis." *Angew. Chem, 52,1:* 131 – 146. https://doi.org/10.1002/anie.201207081.

[37] Nicolaou, K.C., & Montagnon, T. (2008). *Molecules That Changed the World*, 366: Wiley-VCH, Weinheim. https://www.wiley.com/en-us/Molecules+That+Changed+the+World+-p-9783527309832.

[38] Wess, G., Urmann, M., & Sickenberger, B. (2001). "Medicinal Chemistry: Challenges and Opportunities." *113: 3443-3453. Angew. Chem.,* https://doi.org/10.1002/1521-3773(20010917)40:18<3341::AID-ANIE3341>3.0.CO;2-D.

[39] Lipinski, C.A., Lombardo, F., Dominy, B.W., & Feeney, P.J. (1997). "Experimental and Computational Approaches to Estimate Solubility and Permeability in Drug Discovery and Development Settings." *Adv. Drug Delivery Rev.*, 23: 3- 25. http://dx.doi.org/10.1016/S0169-409X(96)00423-1.

[40] Lipinski, C.A. (2004). "Lead- and Drug-like Compounds: the rule of five revolutions." *Drug Discovery Today Technol,* 1, 4: 337 – 341. DOI: 10.1016/j.ddtec.2004.11.007.

[41] Overington, J.P., Al-Lazikani, B., & Hopkins, A.L. (2006). "How many drug targets are there?" *Nat Rev Drug Discov.*, 5, 12: 993-996. Toxicity of drugs- book chapter-New.docx.

[42] Leeson, P.D., & Springthorpe, B. (2007). "The influence of drug-like concepts on decision making in medicinal chemistry." *Nature Reviews Drug Discovery,* 6: 881 – 890. https://doi.org/10.1038/nrd2445.

[43] Hughes, J.D., Blagg, J., Price, D.A., Bailey, S., DeCrescenzo, G.A., Devraj, R.V., Ellsworth, E., Fobian, Y.M., Gibbs, M.E., Gilles, R.W., Greene, N., Huang, E., Krieger-Burke, T., Loesel, J., Wager, T., Whiteley, L., & Zhang, Y. (2008). "Physiochemical drug properties associated with in vivo toxicological outcomes." *Bioorg. Med. Chem. Lett.*, 18, 17: 4872 – 4875. DOI: 10.1016/j.bmcl.2008.07.071.

[44] Ritchie, T.J., & Macdonald, S.F.J. (2009). "The impact of aromatic ring count on compound developability-are too many aromatic rings a liability in drug design?" *Drug Discovery Today,* 14: 1011–1020. Toxicity of drugs- book chapter-New.docx.

[45] Lovering, F., Bikker, J., & Humblet, C. (2009). "Escape from Flatland: Increasing saturations as an approach to improving clinical success." *J. Med. Chem.*, 52, 21: 6752 – 6756. https://doi.org/10.1021/jm901241e.

[46] Paul, S.M., Mytelka, D.S., Dunwiddie, C.T., Persinger, C.C., Munos, B.H., Lindborg, S.R., & Schacht, A.L. (2010). "How to Improve R&D Productivity: The Pharmaceutical Industry's Grand Challenge." *Nat. Rev. Drug Discovery,* 9: 203 – 214. https://doi.org/10.1038/nrd3078.

[47] Altshuler, J.S., Balogh, E., Barker, A.D., Eck, S.L., Friend, S.H., Ginsburg, G.S., Herbst, R.S., Nass, S.J., Streeter, C.M., & Wagner, J.A. (2010). "Opening up to pre-competitive collaboration." *Sci. Transl. Med.*, 2, 52: 52cm26. DOI: 10.1126/scitranslmed.3001515.

[48] Roughley, S.D., & Jordan, A.M. (2011). "The Medicinal Chemist's Toolbox: An Analysis of Reactions Used in the Pursuit of Drug Candidates." *J. Med. Chem.,* 54, 10: 3451 – 3479. https://doi.org/10.1021/jm200187y.

[49] Stepan, A.F., Walker, D.P., Bauman, J., Price, D.A., Baillie, T.A., Kalgutkar, A.S., & Aleo, M.D. (2011). "Structural alert/reactive metabolite concept as applied in medicinal chemistry to mitigate the risk of idiosyncratic drug toxicity: a perspective based on the critical examination of trends in the top 200 drugs marketed in the United States." *Chem. Res. Toxicol,* 24, 9: 1345 – 1410. DOI: 10.1021/tx200168d.

[50] Walters, W.P., Green, J., Weiss, J.R., & Murcko, M.A. (2011). "What do medicinal chemists actually make? A 50-year retrospective." *J. Med. Chem.,* 54, 19: 6405 – 6416. DOI: 10.1021/jm200504p.

[51] Rask-Andersen, M., Almn, M.S., & Schiçth, H.B. (2011). "Trends in exploitation of novel drug targets." *Nat Rev Drug Discov.*, 10, 8: 579 – 590. DOI: 10.1038/nrd3478.

[52] Leeson, P.D., & St-Gallay, S.A. (2011). "The influence of the organizational factor on compound quality in drug delivery." *Nature Review Drug Discovery,* 10: 749 – 765. Toxicity of drugs- book chapter-New.docx.

[53] Arrowsmith, J. (2011). "Trial watch: Phase II failures 2008-2010." *Nat. Rev. Drug Discovery,* 10, 5: 328 – 329. DOI: 10.1038/nrd3439.

[54] Orchard, S., Al-Lazikani, B., Bryant, S., Clark, D., Calder, E., Dix, I., Engkvist, O., Forster, M., Gaulton, A., Gilson, M., Glen, R., Grigorov, M., Hammond-Kosack, K., Harland, L., Hopkins, A., Larminie, C., Lynch, N., Mann, R.K., Murray-Rust, P., Piparo, E.L., Southan, C., Steinbeck, C., Wishart, D., Hermjakob, H., Overington, J., & Thornton, J. (2011). "Shouldn't enantiomeric purity be included in the minimum information about a bioactive entity? Response from the miabe group." *Nat. Rev. Drug Discovery*, 10: 661-669. https://doi.org/10.1038/nrd3503-c2.

[55] Sellers, W.R. (2011). "A blueprint for advanced genetic-based cancer therapy." *Cell,* 147, 1: 26- 31. https://doi.org/10.1016/j.cell.2011.09.016.

[56] Palma, M.D., & Hanahan, D. (2012). "The biology of personalized cancer medicine: facing individual complexities underlying hallmark capabilities." *Mol. Oncol.*, 6, 2: 111 – 127. doi: 10.1016/j.molonc.2012.01.011.

[57] Yarchoan, R., Klecker, R., Weinhold, K., Markham,.P., Lyerly, H., Durack, D., Gelmann, E., Lehrman, S., Blum, R., & Barry, D. (1986). "Administration of 3'-azido-3'-deoxythymidine, an inhibitor of HTLV-III/LAV replication, to patients with AIDS or AIDS-related complex." *Lancet.*, 1, 8481: 575–80. doi: 10.1016/S0140-6736(86)92808-4.

[58] Mitsuya, H., Weinhold, K., Furman, P., St. Clair, M., Li, L., Lehrman, S., Gallo, R., Bolognesi, D., Barry, D., & Broder, S. (1985). "3'-Azido-3'-deoxythymidine (BW A509U): an antiviral agent that inhibits the infectivity and cytopathic effect of human T-lymphotropic virus type III/lymphadenopathy-associated virus in vitro." *Proc Natl Acad Sci USA*,. 82, 20: 7096–100. doi: 10.1073/PNAS.82.20.7096.

[59] Furman, P., Fyfe, J., St Clair, M., Weinhold, K., Rideout, J., Freeman, G., Lehrman, S., Bolognesi, D., Broder, S., & Mitsuya, H. (1986). "Phosphorylation of 3'-azido-3'-deoxythymidine and selective interaction of the 5'-triphosphate with human immunodeficiency virus reverse transcriptase." *Proc Natl Acad Sci USA*., 83,21: 8333–7. doi: 10.1073/pnas.83.21.8333.

Chapter 3

Pollution by Pharma Industries: An Overview

Neeraj Mohan Gupta[1,*], Jagat Singh Kirar[1], Yogesh Deswal[2], Laxmi Deswal[2] and Kailash Chandra[3]

[1]Department of Chemistry, Govt. P.G. College, Guna, Madhya Pradesh, India
[2]Department of Chemistry, Guru Jambheshwar University of Science and Technology, Hisar, Haryana, India
[3]Department of Chemistry, Bareilly College, Bareilly, Uttar Pradesh, India

Abstract

Pharmaceuticals act as lifesaving tools for living beings, but recently, a new face of pharmaceuticals has been seen in the form of pollution because of improper management of by-products and waste materials released by pharmaceutical industries. The advancements in the development of mass spectrometry and analytical methods during recent years have facilitated studies of pharmaceutical residues and wastes in the environment. Though early evidence of the waste generated by pharma industries that served as a source of active pharmaceutical ingredients (APIs) in the environment was sporadic, these investigations received little attention at that time.

In the late 1990s, the finding of oestrogens in sewage and wastewater as a cause of fish feminization triggered a surge of interest in pharmaceuticals in the environment, particularly the role of excreted medications. Following that, ecotoxicologists took an interest in pharmaceutical waste found in sewage effluents and rivers, which may have diverted attention away from potential alternate sources of pharmaceutical residues. The identification of diclofenac residues in cow

* Corresponding Author's Email: theneerajmohan10@gmail.com.

In: Pharmaceuticals: Boon or Bane
Editor: Neha Agarwal
ISBN: 979-8-88697-487-4

carcasses as a cause of vulture population decline in India taught us that exposure routes are often not predictable. There are many more categories of drugs like antibiotics, analgesics, and their metabolites that cause serious harm to the environment and living beings. Therefore, it has been essential to diagnose and to study the pollution caused by pharma industry on a global level so that potential measures can be taken to minimize the problems caused by pharmaceutical waste. Therefore, steps can be taken to minimize their serious post effects and repercussions and potential strategies can be framed to cope with this global issue.

Keywords: pharma industry, pollution, ecotoxicology, active pharmaceutical ingredients

1. Introduction

Concerns have been raised in recent years regarding the existence, prognosis, and effects of new toxins that are being discharged into the environment. Emerging contaminants (ECs) are synthetically produced organic chemicals that may be present in personal care and hygiene products. Pharmaceuticals, personal care products, endocrine disruptors, insecticides, surfactants, and industrial additives are just a few of the many chemicals that make up ECs. The majority of these chemicals is regarded as recently produced substances and are not yet subject to regulation [1-4]. Our ecosystem has been harmed by the production, consumption, and subsequent release of large amounts of toxic chemicals into the environment [5]. The scientific community is becoming more aware of it and is now focused on the need to look into the prevalence and concentration of new pollutants in the environment [6-8]. It occurs as a result of rising worries about the impact of these compounds on non-target creatures in the environment [9]. Risks include accumulation in soil, plants, and animals, microbial resistance to antibiotics, feminization or masculinization of exposed aquatic creatures and interaction with higher organisms' endocrine systems [5, 9-14]. Daughton and Ternes recognized pharmaceuticals as emerging pollutants roughly 20 years ago [15, 16]. Since then, research has been conducted on the global distribution and fate of these pollutants using more sensitive analytical methods in various environmental matrices, such as drinking water, tap water, groundwater, and wastewater effluents [16-19].

Pharmaceuticals are the substances that are used to treat (cure/mitigate), modify or prevent illness, improve health state, or function of the body [15]. Pharmaceuticals are among the substances that are most frequently found in emerging pollutants [20-24]. These compounds may enter the environment through a variety of means, including human excreta, expired medicines, unnecessary storage due to widespread availability of over-the-counter medicines and improper disposal of medicines by hospitals and manufacturers [5, 9, 14, 25-27]. Pharmaceuticals can enter the environment through various comprehensive channels and get accumulated there, as shown in Figure 1.

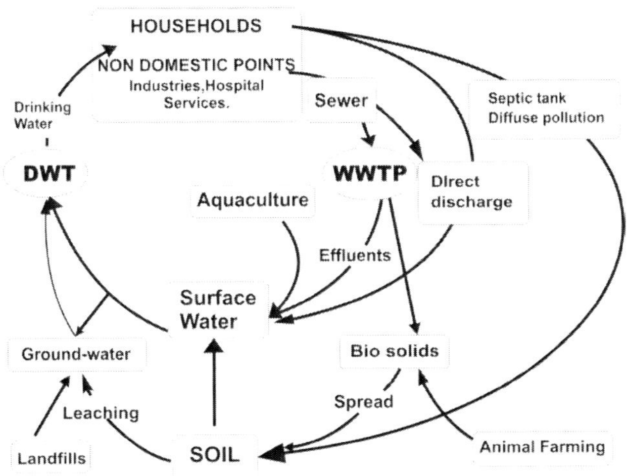

Figure 1. Origin and routes of pharmaceutical products.

Unfortunately, wastewater treatment systems are not designed to remove these non-biodegradable pharmaceutical pollutants from sewage sludge. The majority of conventional wastewater treatment facilities employ primary and secondary treatment processes that cannot get rid of these toxins from water. Because of this, it has been shown that sewage treatment plants are a major entry point for these drugs into water bodies, particularly during outfall discharges into surface water environments [28-31].

2. Occurence/Generation of Pharmaceutical Waste

Numerous pharmacological compounds, including analgesics, antibiotics, and stimulants, have been found in soil and aquatic resources. A single, distinguishable source of pollution, such as a pipe or a drain, is referred to as a point source by the Environmental Protection Authority (EPA) of Victoria [21].This is a typical method of dumping industrial trash into rivers and the ocean. The EPA regulates high-risk point-source waste emissions via the work approval and licencing system, as well as related compliance and enforcement actions. The word "diffuse pollution" is frequently used to describe non-point sources of pollution, which are those inputs and effects that are dispersed over a large area and are difficult to attribute to a single source. Instead of being isolated point sources, they are frequently connected to specific land uses.

According to the European Environmental Agency [32], a large number of activities without a clear point of discharge might result in diffuse pollution. The main source of diffuse pollution is agriculture, although other significant sources include urban land, forestry, air deposition, and rural residences. There are two main ways drugs enter the environment: point source and diffused source [19]. Numerous pharmaceutical compounds or metabolites have recently been found in a variety of aquatic settings and have been documented in the literature [33, 34]. Various sources and the concentrations of different drugs have been summarized in Table1.

Some of the pharmaceutical wastes found in various locations throughout the world are displayed in the table [49]. The table also displays their concentrations, and it is obvious that pharmaceutical waste pollution poses a serious threat to the environment globally. The most important new category of environmental contaminants is the pharmaceutical industry. Sediments, sludge, effluents, natural waterways, and wastewater have all been discovered to contain them. Recent discoveries show their presence in samples collected from all over the world [50]. Examples include the discovery of 12 medicines, in addition to 3 steroid hormones, in 4 WWTPs located in Rome (the most populous city in Italy) in amounts observed in the waters of the Aniene and Tiber rivers [50]. The finding demonstrates that the traditional approach to wastewater treatment does not effectively remove medicines that are intended for removal. It leads to the presence of a variety of pharmaceuticals in known quantities in effluents when samples are taken in both seasons (spring and winter).

Table 1. Pharmaceutical waste sources and occurrences

S. No.	Compounds	Sources	Concentrations	Source Type	Ref.
1	Tylosin	Water system of Umgeni River, KwaZulu-Natal, South Africa.	0.2 1-21.99 µg/L	Diffuse Source	[38]
		Arkansas streams of USA	0.06and 0.18 µg/L	Diffuse Source	[39]
2	Streptomycin	Water system of Umgeni River, KwaZulu-Natal, South Africa.	0.81-8.42 µg/L	Diffuse Source	[38]
3	Sulfamethoxazole	Water system of Umgeni River, KwaZulu-Natal, South Africa.	3.68 µg/L	Diffuse Source	[38]
		Drinking water of USA between 2006 and 2007.	12 ng/L	Point Source	[40]
		River and sewage water.	0.33-0.61 µg/L	Diffuse Source	[41]
		Effluents arising from waste water treatment plants in Scotland.	7 ng L-1	Point Source	[37]
4	Nalidixic acid	Water system of Umgeni River, KwaZulu-Natal, South Africa.	1.73-30.84 µg/L	Diffuse Source	[38]
		Hospital effluent, wastewater treatment plant effluent in surface water, Australia	0.75 µg/L	Point Source	[42]
5	Ampicillin	Water system of Umgeni River, KwaZulu-Natal, South Africa.	2.52 to 14.48 µg/L	Diffuse Source	[38]
		Industrial and wastewater treatment, Madrid (Spain)	2.2-25.6 µg/L	Point Source	[43]
6	Erythromycin	Water system of Umgeni River, KwaZulu-Natal, South Africa.	0.58 - 22.57 µg/L	Diffuse Source	[38]
		Wastewater effluents in Germany	6.0 µg/L	Point Source	[43]
7	Tetracycline	Water system of Umgeni River, KwaZulu-Natal, South Africa.	0.64 .5.68 µg/L	Diffuse Source	[38]
		Wastewater and surface water of China	>1 mg/L waste water,>0.25 mg/L in surface water	Diffuse Source	[44]
8	Ciprofloxacin	Water system of Umgeni River, KwaZulu-Natal, South Africa.	0.7 1 µg/L .16.9 µg/L.	Diffuse Source	[38]
9	Chloramphenicol	Water system of Umgeni River, KwaZulu-Natal, South Africa.	0.5-10.7 µg/L	Diffuse Source	[38]
		industrial and sewage treatment plant Wastewaters in Germany	4.0- 10 µg/L	Point Source	[43]
Lipid regulator					
10	Bezafibrate	Water system of Umgeni River, KwaZulu-Natal, South Africa.	0.81-8.71 µg/L	Diffuse Source	[38]
		Sewage treatment plants and receiving water in Finland	0.42 µg/L	Point Source	[38]
11	Gemfibrozil	Wastewater treatment plants, Rome (Italy)	56–1,032 ng/L	Point Source	[36]

Table 1. (Continued)

S. No.	Compounds	Sources	Concentrations	Source Type	Ref.
β-blocker					
12	Atenolol	Water system of Umgeni River, KwaZulu-Natal, South Africa.	0.96-39.10 µg/L	Diffuse Source	[38]
Antipyretics					
13	Ibuprofen	Water system of Umgeni River, KwaZulu-Natal, South Africa.	0.8 – 18.9 µg/L	Diffuse Source	[38]
		Wastewater in California, USA	3.23-25.8µg/L	Point Source	[47]
		Raw and treated wastewater in Spain	2.3-10.4 µg/L	Point Source	[48]
		Wastewater treatment plants, Rome (Italy)	41-184 ng/L	Point Source	[36]
14	Ketoprofen	Water system of Umgeni River, KwaZulu-Natal, South Africa.	0.4-8.2 µg/L	Diffuse Source	[38]
15	Diclofenac	Water system of Umgeni River, KwaZulu-Natal, South Africa.	1.1-1 5.6 µg/L	Diffuse Source	[38]
		Raw and treated wastewater in Spain	0.48-0.76 µg/L	Point Source	[47]
		Effluents arising from Wastewater treatment plants in Scotland	24.2 - 927 ng L-1	Point Source	[48]
		Wastewater treatment plants, Rome (Italy)	321-1424 ng/L	Point Source	[36]
16	Aspirin	Water system of Umgeni River, KwaZulu-Natal, South Africa.	2.2-10.0µg/L	Diffuse Source	[38]
		Raw and treated wastewater in Spain	2.2-6. I µg/L	Point Source	[48]
17	Acetaminophen	Water system of Umgeni River, KwaZulu-Natal, South Africa.	5.3-58.7 µg/L	Diffuse Source	[38]
		Wastewater influents in France	11.3 µg/L	Point Source	[46]
		Surface water in France	10.6-68.1 ng/L	Diffuse Source	[45]
Anticonvulsants					
18	Carbamazepine	Wastewater treatment plants, Rome (Italy)	69-886 ng/L	Point Source	[36]

3. Adverse Outcomes of Pharmaceutical Pollution

Direct pharmaceutical emissions from the manufacture of medicines can result in extremely high quantities of them in the environment (up to several mg/l), which is a matter of great concern [51]. The "pharmacy of the developing world," India and China, provide the majority of the data that is scarcely available [52]. However, there is other research coming from a number of European nations, including the United States (Larsson, 2014). It will be necessary to invest in effective industrial wastewater technology. However, many nations, including Spain, use a generic substitution system for state-subsidized pharmaceuticals that focuses on cost-cutting and offer no incentive for businesses to invest in "greener" technology. There are several stakeholders with the capacity to develop incentives and encourage more sustainable and innovative attitudes, including regulatory agencies and healthcare organizations that make purchases. As an illustration, the Swedish government now takes factory pollution management into account when businesses compete to get product subsidies [51]. Hospitals, which are often found in metropolitan areas, are a significant source of some particular medications (contrast media, specific broad-spectrum antibiotics, certain cytostatic medicines, etc.). Compared to urban wastewater, hospital wastewater contributes to a lesser extent to the overall mass load of pharmaceuticals in WWTP influents.

Each and every environmental matrix, including seawater, is being impacted by pharmaceuticals. However, despite the fact that 41% of the world's population lives within coastal boundaries, the possible effects of drugs on marine life and coastal settings have got little attention so far [53]. There is proof that medicines have an impact on marine and coastal ecosystems. Most of the information provided so far on pharmacological concentrations in marine species relates to the use of antibiotics in aquaculture [53]. In many locations, the data is scarce or nonexistent (Africa, South America, and Oceania). For instance, research discovered several medications (including coffee and antidepressants like fluoxetine) in sponges from coral reefs in the Maldives archipelago [54]. The presence of medicines in marine biota should be of concern, especially in regions where coral is crucial to the lives and economics of the local population. A recent Italian study revealed that medicines were found throughout the tissues of Mediterranean mussels (carbamazepine was measured in more than 90 percent of the analyzed samples). It is alarming that organisms were found in places that represented relatively pristine marine waters around the Adriatic and Tyrrhenian Seas [55].

Additionally, studies indicate that pharmaceuticals can bioaccumulate through the food web in addition to reaching saltwater and associated water-filtering species. In particular, a study conducted in North-Western Spain revealed the presence of numerous chemicals in the tissues of typical seafood species including bivalves, cephalopods, arthropods, and fish of great economic value. In the tissues of the common octopus, the benzodiazepine alprazolam and the antidepressant citalopram were found [56]. A second study on wild European fish found that for two medications, fish plasma concentrations were greater than the amounts attained in the blood of patients taking the equivalent human medication [57]. Miller et al. study examined all the information on the bioaccumulation of medicines in aquatic life [58]. These studies warn us regarding potential risks related to increased seafood intake in vulnerable demographic groups. But further research is needed to determine how these medications could affect the dynamics of the ecosystem population [59].

Conclusion

In this chapter we have discussed the types of pharmaceutical pollutants arising from various pharmaceutical industries and from hospital waste materials. We have also summarized the effects of pharmaceutical pollutants on aquatic life and on human life. In this chapter, the most significant finding from the literature is how common PPs reach in the aquatic environment. They cannot be eliminated by the sewage treatment technologies now in use. Therefore, it is now necessary to stop additional damage to living things caused by these pollutants. However, the concentrations are often modest; this is particularly true for substances like endocrine-disrupting chemicals that cause substantial, persistent effects at low concentrations. Despite the fact that knowledge is expanding quickly, the issue is not yet fully understood, and there is a severe lack of proper information about risk assessment and management. It would be beneficial to conduct further research on the post effects of different combinations of medicines.

References

[1] Bisarya, S., & Patil, D. (1993). "Determination of salicylic acid and phenol (ppm level) in effluent from aspirin plant." *Res. Ind.*, 38: 170– 172. https://doi/pdf/10.1897/09-073.1.

[2] Holm, J.V., Ruegge, K., Bjerg, P.L., & Christensen, T.H. (1995). "Occurrence and distribution of pharmaceutical organic compounds in the groundwater down gradient of a landfill (Grindsted, Denmark)." *Environ. Sci. Technol.*, 29: 1415– 1420. https://doi.org/10.1021/es00005a039.

[3] Reddersen, K., Heberer, T., & Dunnbier, U. (2002). "Identification and significance of phenazone drugs and their metabolites in ground- and drinking water." *Chemosphere,* 49: 539– 544. https://doi.org/10.1016/s0045-6535(02)00387-9.

[4] Oaks, J.L., Gilbert, M., Virani, M.Z., Watson, R.T., Meteyar, C.U., Rideout, B.A., Shivprasad, H.L., Ahmed, S., Chaudhary, M.J.I., Arshad, M., Mahmood, S., Ali, A., & Khan, A.A. (2004). "Diclofenac residues as the cause of vulture population decline in Pakistan." *Nature*, 427: 630– 633. https://doi.org/10.1038/nature02317.

[5] Ahmed, M.B., Zhou, J.L., Ngo, H.H., Guo, W., Thomaidis, N.S., & Xu., J. (2017). "Progress in the Biological and Chemical Treatment Technologies for Emerging Contaminant Removal from Wastewater: A Critical Review." *Journal of Hazardous Materials*, 323: 274-298. doi: 10.1016/j.jhazmat.2016.04.045.

[6] Agunbiade, F.O., & Moodley, B. (2015). "Occurrence and Distribution Pattern of Acidic Pharmaceuticals in Surface Water, Wastewater, and Sediment of the Msunduzi River, Kwazulu-Natal, South Africa." *Environmental Toxicology and Chemistry*, 31: 36-46. doi: 10.1002/etc.3144.

[7] Brun, G.L., Bernier, M., Losier, R., Doe, K., Jackman, P. and Lee, H.B. (2006) Pharmaceutically Active Compounds in Atlantic Canadian Sewage Treatment Plant Effluents and Receiving Waters, and Potential for Environmental Effects as Measured by Acute and Chronic Aquatic Toxicity. *Environmental Toxicology and Chemistry*, 25, 2163-2176. doi: 10.1897/05-426R.1

[8] Carucci, A., Cappai, G., & Piredda, M. (2006). "Biodegradability and Toxicity of Pharmaceuticals in Biological Wastewater Treatment Plants." *Journal of Environmental Science and Health, Part A*, 41: 1831-1842. https://doi.org/10.1080/10934520600779000.

[9] Benotti, M.J., & Brownawell, B.J. (2009). "Microbial Degradation of Pharmaceuticals in Estuarine and Coastal Seawater." *Environmental Pollution*, 157: 994-1002. doi: 10.1016/j.envpol.2008.10.009.

[10] Belhaj, D., Baccar, R., Jaabiri, I., Bouzid, J., Kallel, M., Ayadi, H., & Zhou, J.L. (2015). "Fate of Selected Estrogenic Hormones in an Urban Sewage Treatment Plant in Tunisia (North Africa)." *Science of the Total Environment*, 505: 154-160. https://doi.org/10.1016/j.scitotenv.2014.10.018.

[11] Ekpeghere, K.I., Lee, J.W., Kim, H.Y., Shin, S.K., & Oh, J.E. (2017). "Determination and Characterization of Pharmaceuticals in Sludge from Municipal and Livestock Wastewater Treatment Plants." *Chemosphere*, 168, 1211-1221. https://doi.org/10.1016/j.chemosphere.2016.10.077.

[12] Gao, L., Shi, Y., Li, W., Niu, H., Liu, J., & Cai, Y. (2012). "Occurrence of Antibiotics in Eight Sewage Treatment Plants in Beijing, China." *Chemosphere,* 86: 665-671. https://doi.org/10.1016/j.chemosphere.2011.11.019.

[13] Huang, J.J., Hu, H.Y., Tang, F., Li, Y., Lu, S.Q., & Lu, Y. (2011). "Inactivation and Reactivation of Antibiotic Resistant Bacteria by Chlorination in Secondary Effluents

of a Municipal Wastewater Treatment Plant." *Water Research*, 45: 2775-2781. https://doi.org/10.1016/j.watres.2011.02.026.

[14] Zhu, Y.G., Johnson, T.A., Su, J.Q., Qiao, M., Guo, G.X., & Stedtfeld, R.D. (2013). "Diverse and Abundant Antibiotic Resistance Genes in Chinese Swine Farms." Proceedings of the *National Academy of Sciences of the United States of America*, 110: 3435-3440. https://doi.org/10.1073/pnas.1222743110.

[15] Daughton, C.G., & Ternes, T.A. (1999). "Pharmaceuticals and Personal Care Products in the Environment: Agents of Subtle Change?" *Environmental Health Perspectives*, 107: 907-938. https://doi.org/10.1289/ehp.99107s690.

[16] Papageorgiou, M., Kosma, C., & Lambropoulou, D. (2016). "Seasonal Occurrence, Removal, Mass Loading and Environmental Risk Assessment of 55 Pharmaceuticals and Personal Care Products in a Municipal Wastewater Treatment Plant in Central Greece." *Science of the Total Environment*, 543: 547-569. https://doi.org/10.1016/j.scitotenv.2015.11.047.

[17] Blair, B.D., Crago, J.P., Hedman, C.J., & Klaper, R.D. (2013). "Pharmaceuticals and Personal Care Products Found in the Great Lakes above Concentrations of Environmental Concern." *Chemosphere*, 93: 2116-2123. https://doi.org/10.1016/j.chemosphere.2013.07.057.

[18] Rubirola, A., Llorca, M., Rodriguez-Mozaz, S., Casas, N., Rodriguez-Roda, I., & Barceló, D. (2014). "Characterization of Metoprolol Biodegradation and Its Transformation Products Generated in Activated Sludge Batch Experiments and in Full Scale WWTPs." *Water Research*, 63: 21-32. https://doi.org/10.1016/j.watres.2014.05.031.

[19] Richmond, E.K., Grace, M.R., Kelly, J.J., Reisinger, A.J., Rosi, E.J., & Walters, D.M. (2017). "Pharmaceuticals and Personal Care Products (PPCPs) Are Ecological Disrupting Compounds (EcoDC)." *Elementa: Science of the Anthropocene*, 5: 66. https://doi.org/10.1525/elementa.252.

[20] Caldas, S.S., Rombaldi, C., De-Oliveira Arias, J.L., Marube, L.C., & Primel, E.G. (2016). "Multi-Residue Method for Determination of 58 Pesticides, Pharmaceuticals and Personal Care Products in Water Using Solvent Demulsification Dispersive Liquid-Liquid Microextraction Combined with Liquid Chromatography-Tandem Mass Spectrometry." *Talanta*, 146: 676-688. https://doi.org/10.1016/j.talanta.2015.06.047.

[21] De, V., Gaffney, J., Almeida, C.M.M., Rodrigues, A., Ferreira, E., Jo-Ao Benoliel, M., & Eepal, S.A. (2015). "Occurrence of Pharmaceuticals in a Water Supply System and Related Human Health Risk Assessment." *Water Res.*, 72: 199-208. DOI: 10.1016/j.watres.2014.10.027.

[22] Deo, R.P. (2014). "Pharmaceuticals in the Surface Water of the USA: A Review." *Current Environmental Health Reports*, 1: 113-122. https://doi.org/10.1007/s40572-014-0015-y.

[23] Ebele, A.J., Abdallah, M.A. E., & Harrad, S. (2017). "Pharmaceuticals and Personal Care Products (PPCPs) in the Freshwater Aquatic Environment." *Emerging Contaminants*, 3: 1-16. https://doi.org/10.1016/j.emcon.2016.12.004.

[24] Madikizela, L.M., Tavengwa, N.T., & Chimuka, L. (2017). "Status of Pharmaceuticals in African Water Bodies: Occurrence, Removal and Analytical

Methods." *Journal of Environmental Management*, 193: 211-220. https://doi.org/10.1016/j.jenvman.2017.02.022.

[25] Daughton, C. (2007). "Green Pharmacy: Preventing Pharmaceutical Pollution." *Journal of Ecology and Toxicology*, 4-13. https://www.scirp.org/(S(i43dyn45te-exjx455qlt3d2q))/reference/referencespapers.aspx?referenceid=2696865.

[26] Riemenschneider, C., Al-Raggad, M., Moeder, M., Seiwert, B., Salameh, E., & Reemtsma, T. (2016). "Pharmaceuticals, Their Metabolites and Other Polar Pollutants in Field-Grown Vegetables Irrigated with Treated Municipal Wastewater." *Journal of Agricultural and Food Chemistry*, 64: 5784-5792. https://doi.org/10.1021/acs.jafc.6b01696.

[27] Sui, Q., Zhao, W., Cao, X., Lu, S., Qiu, Z., Gu, X., & Yu, G. (2017). "Pharmaceuticals and Personal Care Products in the Leachates from a Typical Landfill Reservoir of Municipal Solid Waste in Shanghai, China: Occurrence and Removal by a Full-Scale Membrane Bioreactor." *Journal of Hazardous Materials*, 323: 99-108. https://doi.org/10.1016/j.jhazmat.2016.03.047.

[28] Ekberg, B.M.P., & Pletsch, B.A. (2011). "Pharmaceuticals and Personal Care Products (PPCPs) in the Streams and Aquifers of the Great Miami River Basin." *The Miami Conservancy District.* https://www.mcdwater.org/wp-content/uploads/2017/08/Pharmaceuticals-and-Personal-Care-Products-in-the-streams-and-aquifers-of-the-Great-Miami-River-Basin.pdf.

[29] Focazio, M.J., Kolpin, D.W., Barnes, K.K., Furlong, E.T., Meyer, M.T., Zaugg, S.D., Barber, L.B., & Thurman, M.E. (2008). "A National Reconnaissance for Pharmaceuticals and Other Organic Wastewater Contaminants in the United States-II) Untreated Drinking Water Sources." *Science of the Total Environment,* 402: 201-216. https://doi.org/10.1016/j.scitotenv.2008.02.021.

[30] Kolpin, D.W., Furlong, E.T., Meyer, M.T., Thurman, E.M., Zaugg, S.D., Barber, L.B., & Buxton, H.T. (2002). "Pharmaceuticals, Hormones and Other Organic Wastewater Contaminants in U.S. Streams, 1999-2000: A National Reconnaissance." *Environmental Science & Technology*, 36: 1202-1211. https://doi.org/10.1021/es011055j.

[31] Sgroi, M., Roccaro, P., Korshin, G.V., Greco, V., Sciuto, S., Anumol, T., Snyder, S.A., & Vagliasindi, F.G.A. (2017). "Use of Fluorescence EEM to Monitor the Removal of Emerging Contaminants in Full Scale Wastewater Treatment Plants." *Journal of Hazardous Materials,* 323: 367-376. https://doi.org/10.1016/j.jhazmat.2016.05.035.

[32] de Wilt, A., van Gijn, K., Verhoek, T., Vergnes, A., Hoek, M., Rijnaarts, H., & Langenhoff, A. (2018). "Enhanced Pharmaceutical Removal from Water in a Three Step Bio-Ozone-Bio Process." *Water Research*, 138: 97-105. https://doi.org/10.1016/j.watres.2018.03.028.

[33] Miralles-Cuevas, S., Oller, I., Agüera, A., Llorca, M., Sánchez Pérez, J.A., & Malato, S. (2017). "Combination of Nanofiltration and Ozonation for the Remediation of Real Municipal Wastewater Effluents: Acute and Chronic Toxicity Assessment." *Journal of Hazardous Materials*, 323: 442-451. https://doi.org/10.1016/j.jhazmat.2016.03.013.

[34] Tröger, R., Klöckner, P., Ahrens, L., & Wiberg, K. (2018). "Micropollutants in Drinking Water from Source to Tap-Method Development and Application of a Multiresidue Screening Method." *Science of the Total Environment,* 627: 1404-1432. https://doi.org/10.1016/j.scitotenv.2018.01.27.

[35] Li, J.-D., Cai, Y.-Q., Shi, Y.L., Mou, S.F., & Jiang, G.B. (2007). "Determination of sulfonamide compounds in sewage and river by mixed hemimicelles solid-phase extraction prior to liquid chromatography– spectrophotometry." *Journal of Chromatography A*, 1139: 178–184. https://doi.org/10.1016/j.chroma.2006.11.023.

[36] Patrolecco, L., Capri, S., & Ademollo, N. (2014). "Occurrence of selected pharmaceuticals in the principal sewage treatment plants in Rome (Italy) and in the receiving surface waters." *Environ Sci Pollut Res*, 22: 5864-5876. https://doi.org/10.1007/s11356-014-3765-z.

[37] 37Nebot, C., Falcon, R., Boyd, K.G., & Gibb, S.W. (2015). "Introduction of human pharmaceuticals from wastewater treatment plants into the aquatic environment: a rural perspective." *Environ Sci Pollut Res,* 22: 10559-10568. DOI 10.1007/s11356-015-4234-z. https://doi.org/10.1007/s11356-015-4234-z.

[38] Foluso, O., Moodley, A.B. (2014). "Pharmaceuticals as emerging organic contaminants in Umgeni River water system, KwaZulu-Natal, South Africa." *Environ Monit Assess,* 186: 7273–7291. https://doi.org/10.1007/s10661-014-3926-z.

[39] Haggard, B.E., Galloway, J.M., Green, W.R., & Meyer, M.T. (2006). "Pharmaceuticals and other organic chemicals in selected north-central and northwestern Arkansas streams." *Journal of Environmental Quality*, 35, 4: 1078–1087. https://doi.org/10.2134/jeq2005.0248dup.

[40] Benotti, M.J., Trenholm, R.A., Vanderford, B.J., Holady, J.C., Stanford, B.D., & Snyder, S.A. (2009). "Pharmaceuticals and endocrine disrupting compounds in U.S. drinking water." *Environmental Science and Technology*, 43: 597–603. https://doi.org/10.1021/es801845a.

[41] Watkinson, A.J., Murby, E.J., Kolpin, D.W., & Costanzo, S.D. (2009). "The occurrence of antibiotics in an urban watershed: from wastewater to drinking water." *Science of the Total Environment*, 407: 2711–2723. https://doi.org/10.1016/j.scitotenv.2008.11.059.

[42] Benito-Pena, E., Partal-Rodera, A.I., Leon Gonzalez, M.E., & Moreno-Bondi, M.C. (2006). "Evaluation of mixed mode solid phase extraction cartridges for the preconcentration of beta-lactam antibiotics in wastewater using liquid chromatography with UV-DAD detection." *Analytica Chimica Acta*, 556: 415–422. http://dx.doi.org/10.1016%2Fj.aca.2005.09.054.

[43] Hirsch, R., Ternes, T.A., Haberer, K., & Kratz, K.L. (1999). "Occurrence of antibiotics in the aquatic environment." *Science of the Total Environment*, 225: 109–118. https://doi.org/10.1016/s0048-9697(98)00337-4.

[44] Li, D., Yu, T., Zhang, Y., Yang, M., Li, Z., Liu, M., & Qi, R. (2010). "Antibiotic resistance characteristics of environmental bacteria from an oxytetracycline productionwastewater treatment plant and the receiving river." *Applied and Environmental Microbiology,* 76, 11: 3444–3451. https://doi.org/10.1128%2FAEM.02964-09.

[45] Lindqvist, N., Tuhkanen, T., & Kronberg, L. (2005). "Occurrence of acidic pharmaceuticals in raw and treated sewages and in receiving waters."*Water Research,* 39: 2219–2228. https://doi.org/10.1016/j.watres.2005.04.003.

[46] Rabiet, M., Togola, A., Brissaud, F., Seidel, J.L., Budzinski, H., & Elbaz-Poulichet, F. (2006). "Consequences of treated water recycling as regards pharmaceuticals and drugs in surface and ground waters of a medium-sized Mediterranean Catchment." *Environmental Science and Technology,* 40,17: 5282–5288. https://doi.org/10.1021/es060528p.

[47] Loraine, G.A., & Pettigrove, M.E. (2006). "Seasonal variations in concentrations of pharmaceuticals and personal care products in drinking water and reclaimed wastewater in Southern California." *Environmental Science and Technology*, 40, 3: 687–695. https://doi.org/10.1021/es051380x.

[48] Rodil, R., Quintana, J.B., López-Mahía, P., Muniategui-Lorenzo, S., & Prada-Rodríguez, D. (2009). "Multi-residue analytical method for the determination of emerging pollutants in water by solid-phase extraction and liquid chromatography–tandem mass spectrometry." *Journal of Chromatography A*, 1216: 2958–2969. https://doi.org/10.1016/j.chroma.2008.09.041.

[49] Pruss, A., Giroult, E., Rushbrook, P. (eds). (1999). "Safe management of wastes from healthcare activities." *World Health Organization*. Geneva. 3-4.

[50] Fatta-Kassinos, D., Meric, S., & Nikolaou, A. (2010). "Pharmaceutical residues in environmental waters and wastewater: current state of knowledge and future research." *Anal Bioanal Chem.*, 399: 251–275. https://doi.org/10.1007/s00216-010-4300-9.

[51] Larsson, D.G. (2014). "Pollution from drug manufacturing: review and perspectives." *Philos. Trans. R. Soc. B.,* 369: 20130571. https://doi.org/10.1098%2Frstb.2013.0571.

[52] Rewari, B.B., Mangadan-Konath, N., & Sharma, M. (2020). "Impact of COVID-19 on the global supply chain of antiretroviral drugs: a rapid survey of Indian manufacturers." *WHO South East Asia J. Public Health.*, 2: 126–133. https://doi.org/10.4103/2224-3151.294306.

[53] Gaw, S., Thomas, K.V., & Hutchinson, T.H. (2014). "Sources, impacts and trends of pharmaceuticals in the marine and coastal environment." *Philos. Trans. R. Soc. Lond. Ser. B Biol. Sci.,* 369: 2013057. https://doi.org/10.1098/rstb.2013.0572.

[54] Rizzi, C., Seveso, D., Galli, P., & Villa, S. (2020). "First record of emerging contaminants in sponges of an inhabited island in the Maldives." *Mar. Pollut. Bull.*, 156: 111273. https://doi.org/10.1016/j.marpolbul.2020.111273.

[55] Mezzelani, M., Fattorinia, D., Gorbia, S., Nigro, M., & Regoli, F. (2020). "Human pharmaceuticals in marine mussels: evidence of sneaky environmental hazard along italian coasts." *Mar. Environ. Res.*, 162: 105137. https://doi.org/10.1016/j.marenvres.2020.105137.

[56] Martínez-Morcillo, S., Rodríguez-Gil, J.L., Fernandez-Rubio, J., Rodríguez-Mozaz, S., Míguez-Santiyan, M.P., Valdes, M.E., Barceló, D., & Valcarcel, Y. (2020). "Presence of pharmaceutical compounds, levels of biochemical biomarkers in seafood tissues and risk assessment for human health: results from a case study in

North-Western Spain." *Int. J. Hyg. Environ. Health*, 223: 10–21. https://doi.org/10.1016/j.ijheh.2019.10.011.

[57] Cerveny, D., Grabic, R., Grabicová, K., Randák, T., Larsson, D.G.J., Johnson, A.C., Jürgens, M.D., Tysklind, M., Lindberg, R.H., & Fick, J. (2021). "Neuroactive drugs and other pharmaceuticals found in blood plasma of wild European fish." *Environ. Int.*, 146: 106188. https://doi.org/10.1016/j.envint.2020.106188.

[58] Miller, T.H., Bury, N.R., Owen, S.F., MacRae, J.I., & Barron, L.P. (2018). "A review of the pharmaceutical exposome in aquatic fauna." *Environ. Pollut.*, 239: 129–146. https://doi.org/10. 1016/j.envpol.2018.04.012.

[59] Saaristo, M., Brodin, T., Balshine, S., Bertram, M.G., Brooks, B.W., Ehlman, S.M., McCallum, E.S., Sih, A., Sundin, J., Wong, B.B.M., & Arnold, K.E. (2018). "Direct and indirect effects of chemical contaminants on the behaviour, ecology and evolution of wildlife." *Proc. Biol. Sci.*, 285: 20181297. https://doi.org/10.1098/rspb.2018.1297.

Chapter 4

Emerging Trends in the Pharmaceutical Waste Generation and Disposal after COVID-19

Nitin Gupta[1], Aradhna Patel[2], Haresh Kalasariya[2] and Virendra Kumar Yadav[3,*]

[1]School of Nanosciences, Central University of Gujarat, Gandhinagar, Gujarat
[2]Microbiology Department, Sankalchand Patel University, Visnagar, Gujarat, India
[3]Department of Biosciences, School of Liberal Arts and Sciences,
Mody University of Science and Technology, Lakshmangarh, Rajasthan, India

Abstract

Every year, a million tonnes of pharmaceutical waste is generated around the globe. The major pharmaceutical products include syringes, masks, PPE kits, unused vaccines, antibiotics, etc. Its production was at a steady rate before the pandemic, but during the pandemic, the demand increased many-fold and so did the production. The widespread use of these pharmaceutical products during the COVID-19 pandemic has created an alarming situation around the world. All our landfills, oceans, and rivers are flooded with pharmaceutical waste. The unused antibiotics after unsafe disposal led to the presence of antibiotics in soil and water, which led to the antibiotic-resistant issue. In the current chapter, the authors suggest a drastic change in pharmacy waste before and after the pandemic and ways to minimize the environmental and health risks.

Keywords: COVID-19, color-codes, hazardous; incineration, chlorination

[*] Corresponding Author's Email: yadava94@gmail.com.

In: Pharmaceuticals: Boon or Bane
Editor: Neha Agarwal
ISBN: 979-8-88697-487-4

1. Introduction

Rapid industrialization and urbanization have led to the improvement of the living standards of human beings, but at the same time, they have led to environmental degradation. Several types of waste are generated every day, like industrial waste that includes coal, fly ash, red mud, and gypsum [1], household waste that includes kitchen waste, incense stick ash [2, 3, 4], biomedical waste that includes PPE kits, syringes, blades, etc., and pharmaceutical waste that includes needles, medicines, etc.

The pharmaceutical and personal care products (PPCPs) industries are among the most serious waste-generating industries [5]. According to the pharmacy and poisons board (PPBs), pharmaceutical waste mainly includes medicines that are expired, cut, split, contaminated, or left unused [6]. Moreover, it also covers pharmaceutical bottles, home-use personal care products, boxes, ampoules, intravenous bags, catheters, syringes, contaminated absorbents and protective garments, vials, masks, and gloves [6]. Pharmaceutical waste is generally categorized into three classes, i.e., non-hazardous waste, hazardous waste, and chemotherapeutic waste. Hazardous waste includes any solid waste that exhibits any characteristics of hazardous waste or it is listed as Environmental Protection Agency (EPA) waste that is mainly corrosive, flammable, and reactive. Every year, the demand for medicines and pharmaceutical products keeps on increasing, so the pharmaceutical industry increases manufacturing of new drugs in addition to the existing pharmaceutical products [7]. Every year a million tonnes of pharmaceutical waste are generated around the globe. As per the organization for economic co-operation and development (OECD, 2020), Columbia has the maximum number of hospitals, followed by Japan, the USA, Mexico, and Korea. At the same time, all these countries are major pharmaceutical waste generators. The complete list of countries with the highest number of hospitals is shown below in Figure 1.

The global population, mainly in Asia pacific is exploding rapidly, which has led to a drastic increase in the demand for not only basic resources but also pharmaceutical products, like vaccines, antibiotics, nutraceuticals, etc. The high demand for these pharmaceutical products has led to an increase in the establishment of the pharma industries, which are the leading cause of pharmaceutical waste. But at the same time, the production of pharmaceutical products is also very important.

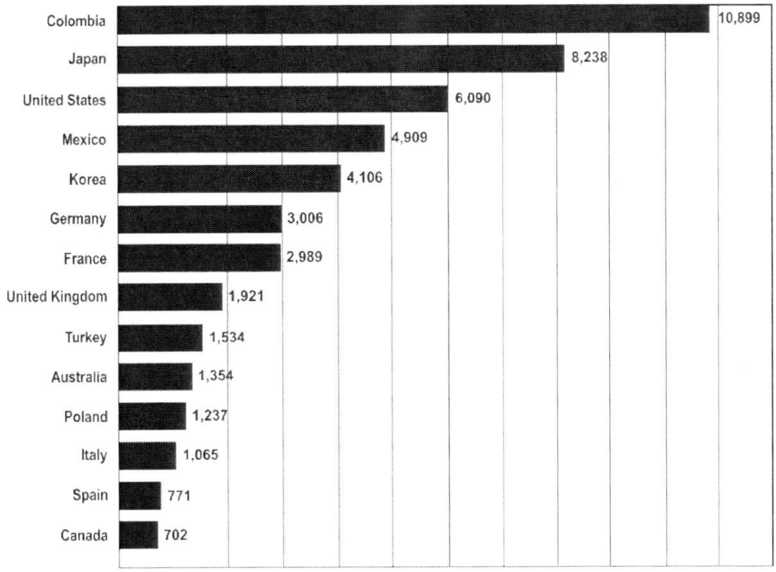

Figure 1. Countries with the highest number of hospitals till 2020 (www. Statistica.com).

The major drawback with such waste is not only environmental hazards, but at the same time, a huge amount of money is required for the disposal of such PPCPs. Due to this disposal, government agencies of any country lose a lot of revenue [8]. A major challenge with the disposal of PPCPs is the lack of awareness among people about the safe handling and disposal of PPCPs, which may result in improper disposal leading to accidental toxicity, increased health costs, antibiotic resistance, landfill pilfering/scavenging, groundwater supply pollution, and ultimately death [9].

Due to the improper disposal of PPCPs from the industries into the effluent system, various types of pharmaceutical waste are present on the land and water. There are several incidences where pharmaceutical waste has been present in the water and soil, from there reaching humans and animals. The two most common pharmaceutical waste products present in our aquatic systems for over a decade is diclofenac and ibuprofen [10]. A study carried out in Serbia, revealed that roughly 74,929 kg of expired drugs are added by households alone every year [11]. Investigators concluded that about 97% of households throw unused pharmaceuticals in the dustbins. A similar study was conducted in South Korea in order to evaluate the environmental risk of pharmaceuticals in wastewater treatment plants. Investigators found several medicine concentrations, for instance, sulfamethoxazole, ibuprofen, acetaminophen, carbamazepine, ketoprofen, and gemfibrozil in nanograms/ liter. The latter three pose a high risk to the fish, while the former two are a potential threat to the green algae [12]. One more study conducted in Germany revealed the presence of about 156 pharmaceuticals on the ground and surface. If this continues, then there is the possibility of the development of drug resistance in humans as well as living beings. During the COVID-19 pandemic, there was a drastic increase in the production of drugs in the world, which has been witnessed due to their frequent use without medical prescriptions by normal healthy individuals [13]. Moreover, during pandemics, people have started sticking to pharmaceutical products for their usage at the time of infection or disease. So, high usage of basic wearables like gloves, aprons, and head caps, led to the generation of the most common pharmaceutical waste generated during the last three years.

In addition to this, there was the blind use of antibiotics and masks, which were used randomly and dumped anywhere in our environment. The disposal of such products has led to a new threat to our environment as these masks might have living pathogens that may affect the living beings in their vicinity. For instance, if this waste is disposed of in the water bodies, then it may affect the aquatic animals. If this waste is disposed of on land, then stray animals like cows and dogs may get exposed to this waste and may get infected or become carriers of several hidden diseases. Similarly, the stocking of antibiotics during this pandemic led to the scarcity of these during the peak time. Once these medicines expire, they get disposed of into the environment. The disposal of such antibiotics into the land or water bodies may lead to antibiotic resistance among the microbes in the soil or water bodies. The development of antibiotic resistance among microbes may affect humans as these antibiotics may stop working for patients [13].

So, there is a need for proper pharmaceutical waste management, which may reduce the chance of diseases in the future. There are several ways to dispose of this pharmaceutical waste, which may reduce the risk of a contagious infection outbreak. For instance, such pharmaceutical waste could be disposed of by using autoclaving, incineration, microwave irradiation, land burial, chemical treatment, etc. [14].

In the current work, the authors have highlighted the drastic increase in the generation of pharmaceutical waste around the globe. Authors have emphasized the proper utilization of pharmaceutical waste as per the directions of authorities. Moreover, the authors have highlighted the various types of pharmaceutical waste in our environment and have proposed emerging techniques for safe disposal.

2. Biomedical Waste and Its Classification

The waste generated from hospitals, during surgeries, room cleaning, etc., is considered biomedical waste. Without proper disposal, it may contain contagious pathogens that may affect other human beings. For the safe disposal of biomedical waste, the government and hospital administration have directives which have to be followed.

2.1. Sources of Biomedical Wastes

There are many sources of generation of bio-medical waste. All of these sources can be divided into two parts, as shown in Figure 2.

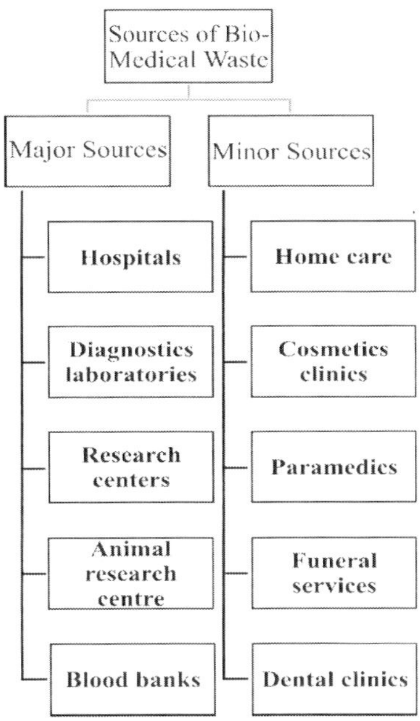

Figure 2. Classification of sources of bio-medical waste; based on hazardous level and quantity.

2.1.1. Primary/Major Sources

The major sources that are responsible for generating bio-medical waste (BMW) include: big hospitals, primary health centers, and other health facilities, diagnostics laboratories and research centers, mortuary and autopsy centers, animal research and testing laboratories, blood banks and other blood collection centers, nursing homes for the elderly. These sources cover most of the parts of BMW and are more hazardous than minor sources [15].

2.1.2. Secondary/Minor Sources

Excluding these sources are several other sources that are also responsible for producing BMW, known as secondary or minor sources of BMW. This category of BMW sources includes home care products, cosmetics & dental clinics, paramedics, funeral services, etc. However, these sources of BMW produce a lower quantity of non-hazardous waste [16].

2.2. Types of Biomedical Waste

BMWs can be divided into two types: hazardous (15-25%) and non-hazardous (75-85%) based on their characteristics, source of generation, and the level of hazardous effects on the environment.

2.2.1. Non-Hazardous Waste

According to the World Health Organization (WHO), various non-hazardous wastes such as wash water, paper cartons, packaging material, food remnants, etc. are responsible for generating 80-85% of the total BMW.

2.2.2. Hazardous Waste

The remaining 15-20% is produced from hazardous waste. Here we are discussing hazardous bio medical waste in detail. Hazardous waste can be further categorized into potentially contagious or infectious waste and potentially toxic non-infectious waste [17]. The infectious hazardous waste contains 15-18% infectious waste and 5-7% toxicity. Toxic waste includes radioactive, chemical, and pharmaceutical waste. The WHO classified biomedical waste into eight categories, which are shown in Figure 3, 1) General waste 2) Pathological waste 3) Radioactive waste 4) Chemical waste 5) Infectious waste 6) Sharps 7) Pharmaceutical waste 8) Pressurized containers.

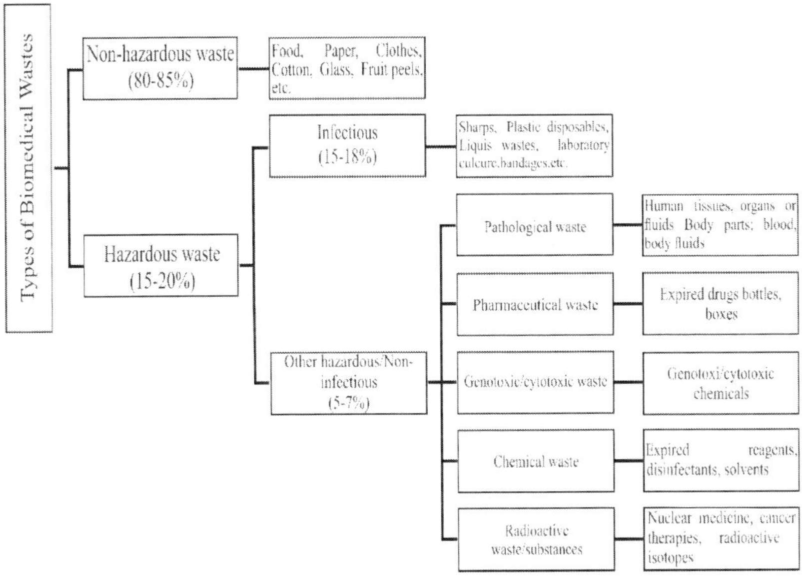

Figure 3. Classification of biomedical wastes based on their characteristics, source of generation, and the level of hazard effect on the environment.

2.2.2.1. Infectious Waste

It is the type of waste that causes infection when a healthy person comes into contact with it. It contains non-sharps, sharps like needles and syringes, plastic disposables, human or animal tissue, blood, and other bodily fluids-soaked bandages, discarded diagnostic samples, laboratory cultures like waste from autopsies and infected animals from laboratories; infectious patient waste like stocks, or swabs, bandages, surgical gloves, disposable medical devices or equipment, etc. that were used to inoculate cultures.

2.2.2.2. Other Hazardous/Non-Infectious Waste

The non-infectious or other hazardous BMWs include various BMWs. The main non-hazardous BMWs are discussed below:

2.2.2.3. Pathological Waste

It contains human tissues, organs, or fluids e.g., body parts; blood and other body fluids; fetuses, and contaminated animal carcasses.

2.2.2.4. Pharmaceutical Waste

It includes expired or no longer needed therapeutic agents (drugs, genes) as well as contaminated pharmaceutical bottles and boxes.

2.2.2.5. Genotoxic/Cytotoxic Waste

It includes genotoxic or cytotoxic substances that are mutagenic, teratogenic, or carcinogenic, such as cytotoxic drugs used in cancer treatment and their metabolites.

2.2.2.6. Chemical Waste

In this type of BMW, waste is generated due to the disposal of solvents and reagents in the environment that have been used for laboratory preparations, disinfectants, sterilant, and heavy metals contained in bio-medical devices (toxic mercury in broken thermometers) and batteries.

2.2.2.7. Radioactive Waste/Substances

Radioactive hazardous waste contains radionuclide-contaminated products, including radioactive diagnostic or radiotherapeutic materials and other radioactive substances, such as left-over liquids that are used in a radioactive research laboratory, contaminated glassware, packages, or absorbent paper, and urine and excreta from patients treated or tested with the uncapped radionuclide [18, 19, 20].

3. Pharmaceutical Waste Sources and Laws

Medical/pharmaceutical waste is being monitored by various laws made by organizational bodies like the Occupational Safety and Health Administration (OSHA), the Centers for Disease Control (CDC), the U.S. Food and Drug Administration (FDA), and the Drug Enforcement Administration (DEA) [21]. For instance, there are generally six sources of pharmaceutical waste; hospitals, pharmaceutical industries, municipal treatment plants, and waste treatment settings like landfills, septic tanks, crop farming, and aquaculture [5].

3.1. Types of Pharmaceutical Waste

Generally, there are various types of risk-associated waste, like chemical waste, pathological waste, pressurized containers, infectious waste, pharmaceutical waste, genotoxic waste, and sharp waste. Chemical waste includes laboratory reagents, disinfectants, and solvents. Pathological waste includes body parts, blood and other body fluids. Pressurized containers generally include waste like gas cylinders, cartridges, and fine aerosol cans. As far as infectious waste is concerned, it mainly includes microbial culture waste from isolation ward tissues and many more. Sharp waste includes needles, infusion sets, scalpels, surgical knives, and blades. Genotoxic waste mainly includes cytotoxic drugs and genotoxic chemicals. Besides this, there is waste with high risks like heavy metal-containing batteries, broken thermometers, and blood pressure gauges. Lastly, there is pharmaceutical waste, which mainly includes expired medicine, unused drugs, and several other items that are not needed anymore. Pharmaceutical waste could be categorized into three categories: non-hazardous, hazardous, and time-sensitive chemo-pharmaceutical waste. Hazardous waste generally includes products which are corrosive, ignitable, and reactive as declared hazardous by EPA. Further hazardous waste could be classified into two types: listed waste and characteristic waste. Listed waste generally includes pharmaceutical products that are used for commercial purposes, whereas characteristic waste exhibits feature like ignitability, corrosiveness, and reactivity [7, 22].

3.2. Features of Pharmaceutical Waste

Along with environmental pollution, the increasing amount of uncontrolled pharmaceutical waste entering the ecosystem has attracted the attention of policymakers to focus on research to reduce harmful emissions and manage pharmaceutical waste. Medicinal waste is synonymous with PPCPs, and according to the World Health Organization (WHO), medical waste is defined as "waste generated in the diagnosis, treatment, or immunization of humans or animals" [22]. Medicines used in the treatment and prevention of diseases in humans and animals are extremely important healthcare tools for society. They are designed to demonstrate specific biological effects in the human or animal body, the evolution of which is still under research. PPCPs could be reactive and have a detrimental effect on the environment and human health. Chemo waste is the most dangerous waste in the medical field. They are highly toxic and corrosive, and they damage natural resources. They are further classified as trace chemotherapy and bulk chemotherapy waste. Trace chemotherapy waste usually includes vials, syringes, IV bags, tubing, gowns, gloves, wipes, and other materials used for regular cleaning and purification of organs. It is now simply referred to as the "Resource Conservation and Recovery Act (RCRA)". Bulk chemotherapy waste includes items that contain chemotherapeutic agents and simply do not qualify as "RCRA". Retired drugs do not pose a serious threat to the environment or other organisms, but their improper disposal can be dangerous if it leads to the contamination of water supplies or local sources used by nearby communities or wildlife. Pharmaceuticals and personal care products (PPCP) have been found to pollute the water and the environment. Pharmaceutical packaging waste is mainly made up of plastics and metals that are difficult to recycle.

4. Color Coding for Storage of Pharmaceutical/ Biomedical Waste

Generally, in the pharmaceutical industry and hospitals, PPCPs are disposed of in various color-coded polybags for safe and proper disposal. As per the biomedical waste rules, 2016, various color coding is provided for the storage of biomedical and pharmaceutical waste prior to their disposal. The government has decided on certain guidelines for the proper disposal of pharmaceutical waste, which have to be followed by all pharmaceutical industries and hospitals. This pharmaceutical waste must be stored as per the policies provided by the government.

As per the directives of the biomedical waste rule 2016, yellow-colored non-chlorinated plastic bags are used for the storage of human anatomical waste, animal anatomical waste, soiled waste, microbiological waste, expired or discarded medicines, micro or bio and other clinical laboratory waste, and chemical liquid waste. Such waste is mainly disposed of either by any one of the following methods, i.e., incinerated or plasma pyrolysis, or deep burial [23].

Red color-coded, non-chlorinated plastic bags or containers are used for the storage of microbiological waste as well as solid waste. This waste generally includes recyclable contaminated waste, for instance, tubing bottles, intravenous tube sets, catheters, urine bags, syringes without needles, and gloves. These are generally treated by auto/micro/hydro and then sent for recycling but not for landfill. For instance, white colored poly bags are containers that are generally translucent, puncture-proof, leak-proof, and tamper-proof are used for the storage of sharp waste, for instance, needles, blades, metals, etc. For the disposal of such wastes, auto or dry heat, sterilization is used, which is followed by shredding or mutilation, or encapsulation. Cardboard boxes with blue-colored markings are used for the disposal of glassware, which has to be disinfected or auto/micro/hydro and then sent for recycling. The black-colored poly plastic bags are used for disposal of discarded waste, incinerated waste, and chemical waste. If all these protocols are followed properly by the pharmaceutical industry, then it becomes very easy for the sanitation workers involved in the disposal of pharmaceutical waste. It also saves time in the segregation of pharmaceutical waste for disposal. It also saves the government a huge amount of money which is being wasted every year on the segregation of pharmaceutical waste before disposal [23].

5. Generation of Pharmaceutical Waste Due to COVID-19

The Indian government has developed an app, named COVID19BWM tracking App, by which one can find out the amount of biomedical waste generated from COVID-19. This app is monitored by the central pollution control board (CPCB, New Delhi). The government has formed COVID-19, biomedical waste generation and treatment facilities, whose main task is to collect, treat, and dispose of the waste. From January to March, 2022, the amount of BMW generated state wise is shown below in Table 1.

The government has monitored the generation and collection of average amounts of COVID-19-relatedBMW during January 2022, February, and March, which is about 38.41 TPD, 33.11 TPD, and 9.87 TPD, respectively. By the end of March 2022, during the pandemic, 93 out of 202 COVID-19 biomedical waste treatment facilities (CBWTFs) started utilizing the Tracking App and nearly 14,054 generators of COVID-19 waste have registered on the COVE 19BAM tracking App (CPCB, 2022).

Table 1. State-wise details of COVID-19 biomedical waste generation in January, February, and March 2022

States/UTs	COVID 19 BMW in January (Tons/day)	No of CBWTFs engaged	COVID 19 BMW in February (Tons/day)	No of CBWTFs engaged	COVID 19 BMW in March (Tons/day)	No of CBWTFs engaged
Andaman and Nicobar	0.00	0	0	0	0.01	0
Andhra Pradesh	1.33	10	1.38	8	0.16	6
Arunachal Pradesh	0.00	0	0.00	0	0	0
Assam	0.12	1	0.12	1	0	0
Bihar	0.2	2	0.15	2	0.04	1
Chandigarh	0.5	1	0.49	0	0.09	1
Chhattisgarh	0.16	3	0.21	3	0.07	3
DD & DNH	0.00	Waste handed to Gujarat facility	0.00	0	0	0
Delhi	3.84	2	2.75		1.02	2
Goa	0.31	0	0.27	0	0.07	0
Gujarat	1.79	13	1.42	10	0.26	9
Haryana	1.75	10	1.28	8	0.58	7
Himachal Pradesh	0.33	2	0.36	1	0.65	2
J & K	0.78	1	0.62	1	0.3	1
Jharkhand	0.33	2	0.07	2	0.02	1
Karnataka	2.5	16	2.00	15	0.74	9
Kerela	7.05	1	8.4	1	2.09	1
Ladakh	0.00				0	0
Lakshadweep	0.00	0	0.00	0	0	0

States/UTs	COVID 19 BMW in January (Tons/day)	No of CBWTFs engaged	COVID 19 BMW in February (Tons/day)	No of CBWTFs engaged	COVID 19 BMW in March (Tons/day)	No of CBWTFs engaged
Madhya Pradesh	0.64	14	1.52	9	0.36	0.36
Maharashtra	5.98		3.47	16	0.81	14
Manipur	0.02	1	0.02	1	0.01	1
Meghalaya	0.08	0	0.11	1	0.05	1
Mizoram	0.16	0	0.18	0	0.09	0
Nagaland	0.00	0	0.00	0	0	0
Odisha	0.98	2	0.76	2	0.18	2
Puducherry	0.32	0	0.34	0	0.05	0
Punjab	0.83	4	0.56	3	0.26	4
Rajasthan	0.40	3	0.42	3	0.11	3
Sikkim	0.08	0	0.09	0	0.03	0
Tamil Nadu	4.81	7	3.5	7	0.97	7
Telangana	0.70	6	0.79	5	0.21	4
Tripura	0.04	0	0.01	0	0	0
Uttarakhand	0.09		0.08	0	0.02	0
Uttar Pradesh	0.93	12	0.74	10	0.28	8
West Bengal	1.36	2	0.99	2	0.34	1
Total	38.41	125	33.11	113	9.87	93

6. Pharmaceutical Waste Treatment and Disposal Methods

As per the government directive, pharmaceutical waste must be disposed of properly; otherwise, it may lead to other consequences for living beings or the environment. Pharmaceutical waste may include pathogens that may cause outbreaks once they come into the environment, so utmost care has to be taken for its disposal. The most important step is to implement government policies, guidelines, and awareness programs for the safe disposal of PPCPs by developing an environment-friendly and sustainable approach. Proper management, treatment, disposal, and segregation of waste are important to minimize waste management and the amount of waste generated is equally important. Less waste means fewer burdens on waste disposal. Therefore, healthcare providers should always try to reduce the production of waste in their daily work in clinics or hospitals [23].

6.1 Autoclaving

An autoclave is an instrument used in the laboratory to kill vegetative cells as well as spores, molds, viruses, etc. The basic principle of the autoclave relies on the fact that the moist heat inactivates the various important microbial enzymes and proteins by coagulating. In autoclaving, one has to attain a temperature of about 121^0C for 15 minutes at 15 psi. At this particular temperature, all the vegetative cells, as well as spores, get killed. Autoclaving uses saturated steam in direct contact with debris in a pressure vessel long enough to kill pathogens. Biomedical waste regulations specify the minimum temperature, pressure, and resistance time for autoclaves for safe disinfection. Autoclaving is not suitable for human anatomy, animals, chemicals, or pharmaceutical waste. This method of sterilizing includes hospital waste, patients' wearables, bedsheets, railways-associated apparel, and hotel rooms. Autoclaving is one of the most important methods for sterilization of interest in hospitals, like surgical tools, blades, bed sheets, etc. For reusing these usable items, they are first kept in autoclave bags or directly in the autoclaves. Some disposable items, like disposable bed sheets, gloves, etc., are autoclaved prior to their disposal into the environment. In order to reduce microbial-associated diseases, the autoclaving method is one of the most reliable techniques in hospitals for the disposal of biomedical and pharmaceutical waste. There are several reports where various investigators have used and suggested autoclaving as one of the most efficient and reliable techniques for the disposal of pharmaceuticals as well as hospital waste. The wider application of the autoclaving method for sterilization of pharmaceutical waste is due to its economic nature, easiness, efficiency, and reliability [24].

6.2. Incineration

Incineration is a method of disposal in which solid organic waste is subjected to high temperatures and dry oxidation so that it can be converted into residues and aerated products. Incineration is the method of burning pharmaceutical waste at a very high temperature, usually 1200^0C under controlled conditions. To attain such a high temperature, a high-power mode of electricity is required, which makes the process more expensive. No doubt, it is one of the most efficient methods for the complete destruction of pathogens associated with pharmaceutical waste, but at the same time, it also costs a lot for government agencies. Every year, a huge amount of money is wasted on the incineration of pharmaceutical waste around the globe. From the literature, it has been found that countries like Bosnia and Croatia, which have efficient incineration facilities and excellent pollution controls, cost around $2.2-4.1 per kg. In the case of the USA, the amount required for the disposal of pharmaceutical waste ranges around$ 4.4-8.2 million whereas in India there is a utilization of 0.5-2.2% of pharmaceutical revenue for the disposal of pharmaceutical waste. For a developing country like India and low-income countries like Kenya and most of the African countries, the incineration technique is used for the disposal of pharmaceutical waste, which is very expensive as most of the revenue goes on the disposal of this pharmaceutical waste [25, 26].

6.3. Microwave Irradiation

Microwave radiation is a method used for the sterilization of pharmaceutical waste from all contagious pathogens. Here, microwaves are used to inactivate pathogenic microbes or cell cultures, et cetera, by exposing them to microwaves. Such a type of sterilization is carried out with specialized equipment, not a nice microwave. Here, pharmaceutical waste stored in various containers is placed inside microwave equipment prior to disposal into the environment. It is also an energy-intensive process that utilizes this electricity and could prove to be an expensive method for occult forums in low-income countries.

6.4. Chemical Disinfection

Chemical disinfection is best suited for the treatment of liquid waste such as blood, urine, feces, or wastewater from a health care facility. The addition of strong oxidants such as chlorine compounds, ammonium salts, or phenolic compounds kills or neutralizes pathogens in the waste. In this method, pharmaceutical waste like PPE kits, masks, apparel, patient bed sheets, gloves, etc. is chemically disinfected with chemicals like sodium hypochlorite, chlorinated water, etc. for the inactivation of living pathogenic micro-organisms or tissues and cells from the hospital waste. Such methods are quite economical in comparison to incineration, autoclaving, and microwave irradiation, as no electricity is required for the disposal of pharmaceutical waste. The disinfectants are spread on the pharmaceutical waste either mechanically or by electronic equipment, which uses less electricity, so the process becomes economical and efficient. Search techniques could become very economical and reliable to find developing as well as low-income countries. But the major drawback with this technique is that after disinfection, the pharmaceutical waste has to be buried and incinerated; otherwise, it will lead to the piling up of the pharmaceutical waste [27].

6.5. Deep Burial

In this method, pharmaceutical waste generated by the hospitals and pharmaceutical industries is generally buried inside the landfills. The pharmaceutical waste must be disinfected by this method as it may contain living pathogens which may reach soil water from another environmental system from the sight of burial. One major drawback of this technique is that it requires a huge amount of land area for the burial of pharmaceutical as well as biomedical waste [28].

6.6. Secure/Safe Land Filling

Safe landfilling involves the disposal of solid waste. Biomedical waste regulations require the disposal of discarded drugs, solid chemical waste, and incinerated ash in protected landfills. In such techniques, large pits are created by the municipalities where pharmaceutical wastes are chemically disinfected before disposal. Further search lands films are covered with soil for a long time. Again, this method is not economical as it requires a huge amount of land area for the landfilling of pharmaceutical as well as medical waste. Such a technique could prove to be cumbersome for a small country that has a very limited area [29, 30, and 31].

6.7. Encapsulation

The method of encapsulation involves stabilizing pharmaceutical products in solid blocks in plastic or steel drums that must be cleaned before use and must not contain previously explosive or hazardous materials. This drum is filled with one-third of its capacity, waste, and the rest of the space is filled with a mixture of cement, cement lime, or bituminous sand. This mixture requires a large amount of water to obtain a paste of the proper liquid consistency. The drums are then sealed and placed at the base of the landfill and covered with municipal waste [32].

6.8. Medical Waste Segregation

In this method, various wastes generated by hospitals and pharmaceutical industries are generally segregated manually or by machines into two separate types of waste. The mechanical segregation involves workers' machines, which makes the process non economical and time-consuming.

7. Role of Pharma Industry and Government Agencies in Minimizing Risk

Both the government and pharmaceutical industries must try to create awareness among the citizens and healthcare workers from time to time about the safe handling and disposal of PPCPs. The agencies must try to create awareness programmes in the form of workshops, seminars, and conferences regularly for all the health line workers. A continuous environmental risk assessment must be conducted by the national drug regulatory bodies arising from the disposal of expired pharmaceuticals. All pharmacists and medical professionals must be provided with proper training on sustainable usage of drugs and proper disposal of pharmaceutical waste at all levels in order to minimize the risk associated with improper disposal of pharmaceuticals. Minimal PPCPs generation at each step of a drug's lifecycle by having an option of taking back, collection at approved sites, and modern technologies for the treatment of wastewater are highly prescribed to minimize the effects of unwanted pharmaceuticals on humans and the environment [33].

Conclusion

The demand for pharmaceutical products is rapidly increasing with time, but its precise and rational use may prevent the generation of unnecessary pharmaceutical waste. The stocking of generic drugs by vendors and citizens may also lead to the expiration of medical drugs. Proper education in the form of workshops and seminars for the health line workers regarding the safe handling and disposal of human waste could prove to be very effective in preventing disease outbreaks. The implementation of an economical and efficient technique for the disposal of pharmaceutical waste could reduce the economic burden on the country's revenue.

References

[1] Yadav, V. K., Fulekar, M. H. (2020). "Advances in methods for recovery of ferrous, alumina, and silica nanoparticles from fly ash waste." *Ceramics.*, 3, 3: 384–420. https://doi.org/10.3390/ceramics3030034.

[2] Yadav, V. K. (2021). "Method for synthesis of amorphous polyhedral shaped iron oxide nanoparticles from incense sticks ash waste." IN *Patent App.* 201811046 769 A.

[3] Yadav, V. K., Gupta, N., Choudhary, N., Singh, J. K., Kumar, G., Inwati Suryaprabha, R. PK. (2020). "Method of processing of incense sticks ash for synthesis of calcium rich zeolites." IN *Patent App.* 202011036844 A.

[4] Gupta, N., Yadav, V. K., Yadav, K. K., Alwetaishi, M., Gnanamoorthy, G., Singh, B., Jeon, B. H., Cabral-Pinto, M. M. S., Choudhary, N., Ali, D., & Nejad, Z. D.

(2022). "Recovery of iron nanominerals from sacred incense sticks ash waste collected from temples by wet and dry magnetic separation method." *Environmental Technology & Innovation,* 25:102150. https://doi.org/10.1016/j.eti.2021.102150.

[5] Fu, Q., Malchi, T., Carter, L. J., Li, H., Gan, J., Chefetz, B. (2019). "Pharmaceutical and Personal Care Products: From Wastewater Treatment into Agro-Food Systems." *Environmental Science & Technology,* 7, 53, 24: 14083–90. https://doi.org/10.1021/acs.est.9b06206.

[6] Nyagah, D. M., Njagi, A., Nyaga, M. N. (2020). "Pharmaceutical waste: overview, management and impact of improper disposal." *Journal of Peer Scientist,* 3, 2: e1000028. doi: 10.5281/zenodo.4088224.

[7] Kadam, A., Patil, S., Patil, S., Tumkur, A. (2016). "Pharmaceutical Waste Management An Overview." *Indian Journal of Pharmacy Practice.*, 9, 1: 2–8. https://www.researchgate.net/publication/303229313_Pharmaceutical_Waste_Management_An_Overview/citation/download.

[8] Ziylan-Yavas, A., Santos, D., Flores, E. M. M., & Ince, N. H. (2022). "Pharmaceuticals and personal care products (PPCPs): Environmental and public health risks." *Environmental Progress & Sustainable Energy*, 41, 4: e13821. https://doi.org/10.1002/ep.13821.

[9] Yang, X., Flowers, R. C., Weinberg, H. S., Singer, P. C. (2011). "Occurrence and removal of pharmaceuticals and personal care products (PPCPs) in an advanced wastewater reclamation plant." *Water Research,* 45, 16: 5218–28. https://www.sciencedirect.com/science/article/abs/pii/S0043135411004234.

[10] Gadipelly, C., Pérez-González, A., Yadav, G. D., Ortiz, I., Ibáñez, R., & Rathod, V. K. (2014). "Pharmaceutical Industry Wastewater: Review of the Technologies for Water Treatment and Reuse." *Industrial & Engineering Chemistry Research*, 53, 29: 11571–92. https://doi.org/10.1021/ie501210j.

[11] Paut Kusturica, M., Tomas, A., Tomic, Z., Bukumiric, D., Corac, A., Horvat, O., & Sabo, A. (2016). "Analysis of expired medications in Serbian households." *Slovenian Journal of Public Health*, 55, 3:195–201. https://doi.org/10.1515/sjph-2016-0025.

[12] Sim, W. J., Lee, J. W., & Oh, J. E. (2010). "Occurrence and fate of pharmaceuticals in wastewater treatment plants and rivers in Korea." *Environmental Pollution,* 158, 5: 1938–47. https://www.sciencedirect.com/science/article/abs/pii/S0269749109005351.

[13] Lai, C. C., Chen, S. Y., Ko, W. C., & Hsueh, P. R. (2021). "Increased antimicrobial resistance during the COVID-19 pandemic." *International Journal of Antimicrobial Agents*, 57, 4: 106324. doi: 10.1016/j.ijantimicag.2021.106324.

[14] Meena, P. B., Dharshini, B., & Gokul, S. (2022). " Biomedical Waste Management Using Incineration and Autoclave." *International Journal for Research in Applied Science and Engineering Technology,* 10, 5: 4645–8. https://doi.org/10.22214/ijraset.2022.43371.

[15] Nwachukwu, N. C., Rodriguez-Morales, O. C. U. E. A. J. (2013). "Health Care Waste Management – Public Health Benefits, and the Need for Effective Environmental Regulatory Surveillance in Federal Republic of Nigeria." In *Current*

Topics in Public Health Edited by Orji FA. Intech Open. https://doi.org/10.5772/53196.

[16] Bio Medical Waste IILS Blog. *Indian Institute of Legal Studies*. https://www.iilsindia.com/blogs/bio-medical-waste/.

[17] Sherani, S. H. (2020). "Proper Hospital Waste Management for the Protection of Human Health and Environment: A Review." *Health Syst Policy Res*, 7, 1: 86. www.imedpub.com.

[18] Harty-Golder, B. (1993). "Biomedical waste." *Journal of the Florida Medical Association*. 80, 3: 205–8. https://pubmed.ncbi.nlm.nih.gov/8487029/#article-details.

[19] Kumari, R., Srivastava, K., Wakhlu, A., & Singh A. (2013). "Establishing biomedical waste management system in Medical University of India – A successful practical approach." *Clinical Epidemiology and Global Health*, 1, 3: 131-136. https://doi.org/10.1016/j.cegh.2012.11.004.

[20] Malsparo-Collection Storage and Treatment of medical waste. (2022). Types of medical waste. Online. https://www.malsparo.com/types.htm.

[21] Shareefdeen, Z. M. (2012). "Medical Waste Management and Control." *Journal of Environmental Protection*, 03, 12: 1625–8. doi: 10.4236/jep.2012.312179.

[22] Magagula, B. K., Rampedi, I. T., & Yessoufou, K. (2022). "Household Pharmaceutical Waste Management Practices in the Johannesburg Area, South Africa." *International Journal of Environmental Research and Public Health*, 19, 12: 7484. https://doi.org/10.3390/ijerph19127484.

[23] Jalal, S. M., Akhter, F., Abdelhafez, A. I., & Alrajeh, A. M. (2021). "Assessment of knowledge, practice and attitude about biomedical waste management among healthcare professionals during COVID-19 crises in Al-Ahsa." *Healthcare (Switzerland)*, 9, 6: 747. doi: 10.3390/healthcare9060747.

[24] Ab, M. (2016). "A Short Review on Pharmaceutical Waste Disposal and Management. Reviews on Pharmaceutics and Nanotechnology Research & Reviews." *Journal of Pharmaceutics and Nanotechnology*, 4. https://www.rroij.com/open-access/a-short-review-on-pharmaceutical-waste-disposal-and-management-.php?aid=82745.

[25] Wajs, J., Bochniak, R., & Golabek, A. (2019). "Proposal of a mobile medical waste incinerator with application of automatic waste feeder and heat recovery system as a novelty in Poland." *Sustainability (Switzerland)*, 11, 18: 4980. doi: 10.3390/su11184980.

[26] Liu, F., Liu, H. Q., Wei, G. X., Zhang, R., Zeng, T. T., & Liu, G. S. (2018). "Characteristics and treatment methods of medical waste incinerator fly ash: A review." *Processes*, 6: 173. doi: 10.3390/pr6100173.

[27] Exner M., Bhattacharya S., Gebel J., Goroncy-Bermes P., Hartemann P., Heeg P., Ilschner C., Kramer A., Ling M. L., Merkens W., Oltmanns P., Pitten F., Rotter M., Schmithausen R. M., Sonntag H. G., Steinhauer K., Trautman M. (2020). "Chemical disinfection in healthcare settings: critical aspects for the development of global strategies." *GMS Hygiene and Infection Control*, 15, doc 36. doi: 10.3205/dgkh000371.

[28] Gaurina-Međimurec, N., Pašić, B., Mijić, P., & Medved, I. (2020). "Deep underground injection of waste from drilling activities—An overview." *Minerals*, 10, 4: 303. https://doi.org/10.3390/min10040303.

[29] Zare, M., Dziemidowicz, K., Williams, G. R., & Ramakrishna, S. (2021). "Encapsulation of pharmaceutical and nutraceutical active ingredients using electrospinning processes." *Nanomaterials*, 11, 8: 1968. https://doi.org/10.3390/nano11081968.

[30] Bu, Q., Cao, H., He, X., Zhang, H., & Yu, G. (2020). "Is Disposal of Unused Pharmaceuticals as Municipal Solid Waste by Landfilling a Good Option? A Case Study in China." *Bulletin of Environmental Contamination and Toxicology*, 105, 5:784–9. https://doi.org/10.1007/s00128-020-03006-5.

[31] Vaverková, M. D. (2019). "Landfill impacts on the environment— review." *Geosciences*, 9, 10: 431. https://doi.org/10.3390/geosciences9100431.

[32] Albaroodi, K. A. I. (2019). "Pharmacists' Knowledge Regarding Drug Disposal in Karbala." *Pharmacy*, 7, 2: 57. doi: 10.3390/pharmacy7020057.

[33] Sung, M., Huang, Y., Duan, Y., Liu, F., Jin, Y., & Zheng, Z. (2021). "Pharmaceutical industry's engagement in the global equitable distribution of COVID-19 vaccines: Corporate social responsibility of EUL vaccine developers." *Vaccines*, 9, 10: 1183. https://doi.org/10.3390/vaccines9101183.

Chapter 5

Critical Risk Assessment of Adverse Effects Associated with Long Term Usage of Various Analgesics on Human Health

Anupma Singh*
Department of Chemistry
DDU Government P.G. College, Sitapur, India

Abstract

Analgesics are compounds that are capable of relieving pain without causing loss of consciousness. Though analgesics are the most commonly used drugs, people lack knowledge of their high-risk use and their association with pain sensitivity. Heavy use of over-the-counter analgesics has been associated with persistent renal failure. Patients with chronic pain face difficulty because of challenges in obtaining adequate care, resulting in physical and emotional stress. The pharmaceutical companies have given us a few options that include non-opioid analgesics, paracetamol, NSAIDs (non-steroidal anti-inflammatory drugs), metamizole and in severe pain, drugs having opioid formulations are also available. A combination of these medicines (opioids and non-opioids) is also given in some chronic cases of pain. The choice of medication should be based on the pain diagnosis and the mechanisms of pain following history, physical examination, and a risk-benefit assessment in such a manner that benefits outweigh the risks. However, each of the above categories of drugs has certain negative effects associated with them. This chapter analyzes the adverse effects associated with heavy and long-term usage of various analgesics in chronic pain conditions, with a special reference to the analgesics used

* Corresponding Author's Email: dranupma0106@gmail.com.

In: Pharmaceuticals: Boon or Bane
Editor: Neha Agarwal
ISBN: 979-8-88697-487-4

by senior citizens with common diseases like arthritis. The goal of this chapter is to ensure best clinical outcomes to protect public health and to limit the adverse outcomes of long-term and heavy use of analgesics.

Keywords: analgesics, painkiller, diagnosis, chronic pain, NSAIDs

1. Introduction

Pain is a complicated experience which is approached by different medicines and influences the quality of life. Even though a wide range of medicines are available for pain, sometimes the pain management is inadequate. Analgesics are the most widely used primary medicines to treat pain and must be used under proper medical consultation. Global data suggests that a majority of patients fail to consume medicines as directed by the doctor; instead, over half of the prescribed medicines are sold and consumed inappropriately [1]. Analgesics and antipyretics are the most commonly used classes of drugs in the world, available with or without a prescription in mono- and multi-component preparations, but each class has its own set of advantages and disadvantages. For example, paracetamol (also known as 4-hydroxy acetanilide or acetaminophen) is a widely used over-the-counter analgesic and antipyretic drug for relieving pain and reducing fever. It is commonly detected in sewage treatment plant effluents, the surface, and drinking water [2], and heavy use of paracetamol is associated with renal failure and other health problems. Until recently, the use of paracetamol in pregnancy has been considered safe. However, the latest available literature shows that it may have negative long-term effects in the form of attention deficit hyperactivity disorder (ADHD) and autism spectrum disorder (ASD) in the offspring if used by the mother during pregnancy [3].

Unfortunately, inadequate pain management leads to chronic and intolerable pain, which is a major healthcare issue that can be considered as a symptom of a disease. Despite the difficulty in classifying and understanding various mechanisms of pain, certain pain syndromes are pervasive. For example, lower back pain, cancer pain, headaches, post-operative pain, and joint pain are some of the most common reasons why patients seek analgesics. This chapter analyzes the potential adverse drug effects (ADEs) associated with commonly used analgesics and suggests ways in which the use of analgesics can be optimized [4].

2. Understanding the Use of Analgesics

Analgesics are substances that work in different ways to relieve different types of pain experienced in the body. Over-the-counter analgesics that are commonly used by the public are paracetamol, weak opioids (codeine), and NSAIDs (ibuprofen and aspirin). Painkillers are divided into three broad categories, namely non-opioids, non-steroidals, and opioids. Each category is described in detail in the following section.

2.1. Non Opioids Painkiller-Acetaminophen

Acetaminophen is also called a non-aspirin pain reliever and an anti-pyretic drug. Unlike NSAIDs, it has no anti-inflammatory action [5]. Acetaminophen being analgesic and antipyretic relieves us from fevers, headaches, and other common pains, but it does not relieve us from inflammation. Acetaminophen is often recommended for arthritis patients because it has fewer side effects than other pain reliever medicines. It is also available in many other forms, such as tylenol, paracetamol, and panadol. It has many side effects for arthritis patients, as these patients require a consistent source of acetaminophen to subside the severe pain in their joints. According to current guidelines, acetaminophen is recommended as a first-line analgesic for mild to moderate pain due to arthritis of the knee and hip [6-12]. The analgesic activity of acetaminophen is associated with inhibiting the synthesis of prostaglandin. However, the primary mechanism of prostaglandin synthesis inhibition by acetaminophen remains unknown [13].

2.1.1. A Study of Acetaminophen Hepatotoxicity

Quite recently, there has been an increasing concern about acetaminophen hepatotoxicity. Due to these concerns, the FDA (Food and Drug Administration) of various countries advised the workers of the "Center for Drug Evaluation and Research" to recommend the dosage reduction of acetaminophen to reduce liver toxicity [14]. An assembly of three advisory people reviewed the report of the working group and gave the following three recommendations:

a. A reduction in the dosage from 4 gm to possibly 3250 mg daily has been prescribed.

b. A ban on the dosage of narcotic-acetaminophen combinations.
c. A reduction has to be made for nonprescription dosage from 1 gm to 650 mg.

Though the FDA did not implement all the recommendations of the advisory, they increased the number of warnings on hepatotoxicity of non-prescription dosages of acetaminophen. They advised limiting the acetaminophen component in a combination of analgesic prescription products to 325 mg per dosage, also requiring a black box warning stating liver injury, and stating one concentration in acetaminophen liquid (160 mg/5ml). The question still remains unanswered as to whether the data will support more risk of hepatotoxicity in older adults or not [14]. Acetaminophen overdose results in liver failure and is dosage-dependent. In healthy adults, the pathways used are glucuronidation and sulfation [15]. However, the toxic metabolite, N-acetyl-p-benzoquinone (NAPQI), binds to glutathione in phase II, shifting the metabolism to phase I because of the saturation after an acetaminophen overdose. NAPQI accumulates and binds to the hepatic cells when glutathione is depleted, causing hepatic necrosis. Recent studies show that similar to the brain, the spinal dorsal horn is critical to pain pathways and modulates nociceptive transmission and acetamino-phen induced analgesia by acting not only on the brain but also the spinal cord [16], which is the main cause or liver failure [17]. In people who are living a malnourished life, the therapeutic doses of acetaminophen cause transient liver enzyme increases in secretions leading to hepatotoxicity [15].

Age-related studies have also been done for acetaminophen hepatotoxicity. For example, in a study, adults aged 18 to 55 years and senior citizens aged more than 70 years were administered to see the impact of acetaminophen hepatotoxicity in a short period of time [18]. The treatment group under observation was given 3 to 4 grams of acetaminophen per day, whereas the control group patients were administered no acetaminophen. At baseline, plasma alanine aminotransferase (ALT) concentrations were seen, and then a random acetaminophen concentration was calculated on the fifth day. The results were astonishing! The younger ones had higher ALT elevations compared to older patients, though they had higher mean acetaminophen concentrations on the fifth day of treatment [18].

In another study, approximately 600 patients aged 17 to 76 years were chosen from different hospitals. It was observed that most of the cases of acute liver failure were due to unintentional overdoses of acetaminophen [19]. The dosage administered to them was 7.5 gram of acetaminophen, 64% were given a combination of opioid and acetaminophen, whereas 38% of cases used multiple acetaminophen products [19]. Another study of emergency visits in numerous hospitals between the years 1993 and 2007 showed that acetaminophen overdose was given to 0.05% of all visits [20]. The lower rate of acetaminophen overdose that was treated in emergency hospitals suggested that older adults were at a lower risk of acetaminophen overdose. The studies also reported that senior citizens may not be taken to hospitals to treat acetaminophen toxicity, and some cases may go unreported due to many pharmacies being involved. Overdose toxicity was most prevalent in the individuals of age 15 to 24 years, according to a survey of National Electronic Injury Surveillance System (NEISS) data from 2006-2007, and the overdose issue was 2.2 percent for senior citizens [21].

2.1.2. A Study of Acetaminophen Pharmacokinetics as Related to Age

Many studies in healthy senior citizens have revealed the pharmacokinetics of acetaminophen [22-27]. It has been shown that the drug absorbance is independent of age and the acetaminophen is successfully absorbed from the gastrointestinal (GI) tract [28]. It has been seen that in females the volume of distribution decreases, which is consistent with the drug's hydrophilic nature and also related to changes in body composition. However, no such marked difference can be observed between a healthy and a weak-healthy senior citizen [24]. However, some research explains the difference in clearance in senior citizens. For example, Wynne et al. showed that there are chances of clearance of glucuronide and sulphate metabolites remaining unaffected in the case of weaker senior citizens [24]. Taking a step further, intravenous exposure of acetaminophen for patients in the age group of 81-91 years had 1.4 to 1.6-fold greater metabolites than patients in the age group of 21–41 years [27, 28]. As is evident now, it is shown that if senior citizens are healthy, the conjugative activity of the liver is variable, but weaker senior citizens do have a danger of acetaminophen toxicity. Therefore, it is suggested that the dosage should be administered in accordance with the age and health.

2.2. Non–Steroidal Anti- Inflammatory Drugs (NSAIDs)

NSAIDs are antipyretics, produce analgesic effects, and reduce inflammation. NSAIDs are primarily used as painkillers for gastric peptic ulcers [29]. The risks associated with popular NSAIDs like Ibuprofen and naproxen [30]; need to be studied deeply due to the risk of many side effects associated with them.

When the person is not in generally good health and the dosage is to be taken for a long time, the risk of side effects increases.

2.2.1. Probable Side Effects of NSAIDs Include

- indigestion – it results in a feeling of sickness and stomach aches.
- stomach ulcers – the lining of the stomach causes it to erode. Therefore, some medication can be taken along with it.
- a throbbing headache
- drowsiness and dizziness
- some allergic reactions
- in some cases, problems with the liver, kidneys, heart, and blood circulation, such as heart failure, heart attacks, and strokes

2.2.2 Interactions of NSAIDs with Other Kinds of Medicine

Some NSAIDs can react unpredictably with other medicines. This can, however, affect how well the medicine works and increase the risk of side effects. It is extremely important to get medical advice before taking an NSAID if a person is already consuming other NSAIDs such as low-dosage of diuretics, aspirin, cyclosporine, warfarin, lithium, methotrexate, citalopram, and fluoxetine (Prozac). When acetaminophen is unable to provide relief, then NSAIDs are given to adults [31, 32]. NSAIDs and other salicylates work by inhibiting the COX-1 and COX-2 (Cyclooxygenase) pathways, which prevents the release of prostaglandins and other pain-causing-secretions [33].

2.2.3. A Study of NSAID Pharmacokinetics as Related to Age

Many people have researched NSAIDs' adverse effects. Several of them have reported that the compounds metabolized by isoenzymes are called Phase I cytochrome P450 [34]. Generally, plasma protein binds NSAIDs well and absorbs them easily. It has also been discovered that elderly people with hypoalbuminemia consume a large concentration of these drugs [37, 38].

2.2.4. Adverse Effects of NSAIDs on Cerebrovascular or Cardiovascular System

Rahme et al. have shown that people with great myocardial infarction (MI) are administered with celecoxib and aspirin as compared to acetaminophen alone in senior adults in Québec [39, 40]. In contrast, Lévesque et al. conducted a retrospective study in Québec (Canada) on 113,937 senior citizens who did not have a myocardial infarction history and began taking NSAIDs to determine whether NSAIDs increased the risk of a first MI case [37]. These studies reflect that celecoxib has no relation to risk being associated with first MI. Another study found that celecoxib and medication, as well as duration of the treatment, had not affected the risk of first MI [38]. Then a study to see the effect of NSAIDs on the rate of heart failure was conducted by Mamdani et al. [39]. They used a sample in which patients were given celecoxib and non-selective NSAIDs to see the relationship between their use and the rate of heart failure. Further research into use of diclofenac found that it was associated with an increased risk of stroke [40]. The relative effects of non-selective and COX-2 selective NSAIDs on a composite cardiovascular outcome showed a relation to heart failure, stroke, and MI [41].

2.2.5. Renal Adverse Drug Effect Caused by NSAID

Schneider et al. took a population sample from Quebec of senior citizens to study the relationship between renal failure and a selective NSAID and non-selective COX-2 [42]. The scientists found that there was a huge risk associated with renal failure within 29 to 30 days of hospitalization; however, the ratio did decrease with time. Moreover, when NSAIDs were administered, the risk of acute renal failure increased even more [42]. The scientists researched more on the conditions that are responsible for acute kidney issues in a sample of senior citizens in Calgary when they were given non-selective and COX-2 as well as other important NSAIDs [43]. The findings were great and showed a decreased rate in estimated glomerular filtration [43].

2.2.6. Other Adverse Drug Effects Associated with NSAIDs

When Kerr et al. studied the mortality rate of senior citizens on medication, they found that the individuals who were given non-selective NSAIDs were prone to a greater risk of death in comparison to those who were under the medication of non-selective COX -2 suffering from ailments like glaucoma and hypothyroidism [44].The individual's heartbeat rates after administering various types of NSAIDs are given in table 1[39].

Table 1. Heartbeat rates after being administered various types of NSAIDs

S. No.	Name of medicine	Range of heartbeat	Mean
1.	Celecoxib	1.26- 1.79	1.38
2.	Meloxicam	1.26- 1.79	1.48
3.	Diclofenac	1.29- 1.63	1.45

The above study of NSAIDs in relation to cardiac rate is given, and it is seen that the risk of mortality increases manifolds. The study used a time-to-event approach, which tells us mortality risk only when the individual is exposed to a drug. It does not cover the aftereffects of the drug being administered. The research was also limited to the use of only non-selective NSAIDs that were administered to one group. These NSAIDs definitely increase the risk of many diseases in senior citizens, like heart problems, brain problems, renal as well as stomach-associated diseases[45].Whenever individuals are given NSAIDs, often when pain is not controlled by acetaminophen, they are required to be monitored regularly since the risk involved is high. Second, before administering NSAIDs to patients, doctors should try to treat them first with other non-acetylated salicylates because they are better medicines for the prevention of minor problems related to the stomach [46].

3. Opioids

If an individual is unable to get relief from pain by using acetaminophen or NSAIDs, then analgesics like opioids are used [46, 47]. Many studies have been conducted to study the effects of opioids' reactions, and pharmacokinetics is typically used to determine the efficiency of the drug. Many other factors are also taken into consideration, like the age of the individual, the length of time from which they have been suffering, and also the intensity of the pain. Opioids react in such a way that they can inhibit pain pathways by binding to various receptors found in our central nervous system [48, 49].

Medications in which opioids bind to receptors are increasingly being prescribed for the treatment of multiple and diverse chronic pain conditions. Their use for extreme pain or terminal pain is well accepted. Their usage in the long-term treatment of chronic noncancerous pain can be controversial for many reasons. One of the primary reasons is the well-known problem of psychological addiction that can occur with the use of these medications. Abuse and diversion of these medications is an increasing problem as their availability increases, and this public health issue confounds their clinical utility. Also, the extent of their efficiency in the treatment of pain when utilized for a chronic condition has not been definitively proven. The usage of opioids in the treatment of chronic pain is also influenced by the fact that these potent analgesics are associated with a notably huge number of side effects and complications. Common side effects of opioid medicines include physical addiction, sedation, dizziness, respiratory problems, nausea, constipation, and vomiting. Physical addiction is the most important clinical concern that may prevent proper prescribing and, in turn, inadequate pain management. Other common side effects include hormonal dysfunction, delayed gastric emptying, muscle rigidity, hyperalgesia, immunologic dysfunction, and myoclonus. The most common side effects of opioid use as medicine are constipation (which has a very high incidence) and nausea. These two severe side effects can be difficult to manage, and tolerance to them does not develop, specifically for constipation. These conditions are severe enough to require opioid discontinuation and contribute to under-dosing and inadequate analgesia. Several clinical trials are ongoing to identify adjunct therapies that may mitigate these side effects. Switching opioids and/or routes of administration of these medicines may also provide benefits for patients. Proper patient screening, preemptive treatment, and education of potential side effects may help in maximizing effectiveness while reducing the severity of side effects and adverse events. Opioids are usually considered broad-spectrum analgesic agents, affecting a wide number of organ systems and influencing a large number of body functions [50, 51].

3.1. A Study of Opioids Pharmacokinetics in Relation to Age

Many studies show that when opioids are used, there is a reduction in removal of oxycodone. However, this is mostly related to age [52-54]. Scientists also studied the pharmacokinetics of fentanyl and tramadol in senior citizens which showed that in these drugs the effect of removal is based on age [55, 56]. The CYP2D6 is in charge for converting prodrugs such as tramadol and codeine into active metabolites [56, 57]. The effect of parent drug like meperidine has a tendency to accumulate in senior citizens that may result in neurotoxicity, so doctors usually avoid its usage [58]. In senior citizens it has been seen in case of methadone which has a very long half life and also some potential QT interval effects associated with it. Therefore doctors generally do not recommend it for these individuals [59]. The drug pentazocine is not given to senior citizens because of its pharmacodynamic reactions, and also because some changes have been observed in its pharmacokinetics which are related to age.

3.2. Adverse Effects of Drug-Delirium in Opioid-Treated Patients

A study was conducted in 2011 to assess the risk of drug delirium in senior citizens [60]. It was observed that those individuals who had opioid medication were at a risk that was double that of those without opioid medication for delirium [61].

3.3. Adverse Drug Effects in Opioid-Treated Patients

It was seen that a group of scientists, Takkouche et al., took the data from six different research projects and published it after deep analysis [62]. In their analysis, they found that the individuals who were given opioids were 39% more prone to fractures than the individuals without them. Miller et al. investigated the effects of opioid treatment on hip and wrist fractures, as well as attempted to conduct a comparative study for the NSAIDs in senior citizens having arthritis [63]. The research group of Saunders et al. also tried to calculate the risk associated with senior citizens' fractures who were on opioid medication [64]. Keeping many variables constant, it has been discovered that opioid medication increases the risk of fracture by five times as compared to NSAIDs. Solomon's group studied the pain of senior citizens and compared the risk of oral opioids like codeine or tramadol to hydrocodone. Further, their studies concluded that patients who were taking tramadol were at a much higher risk than the individuals who were given hydrocodone medication [65].

A group of scientists named Buckeridge from Canada varied the time exposed to opioids and also grouped them by their potency research to observe the risk involved with injuries to softer tissues [66]. They found that individuals who were exposed to low potency opioids like codeine were at a much lower risk of injuries and fractures. The other lower-potency drugs that can be used are butorphanol, pentazocine, or oxycodone. Many scientists also took into consideration various CNS active medicines and other problems like gait and balance issues in senior citizens. It has been seen that individuals under the treatment of these CNS active medicines that may range from antidepressants, antipsychotics are more prone to falls and damage to their bodies than people who have not been under such treatments [67]. Another surprising study that was carried out by a group of scientists led by Dublin found that individuals who were using opioid medication were more prone to getting the infection of pneumonia [68].

3.4. Other Adverse Drug Effects of Opioid-Addicted Patients

Researchers showed that individuals treated with some specific immuno-suppressive opioids like morphine, codeine, methadone or fentanyl had a greater risk of pneumonia. This risk has further been aggravated by the presence of those using the sustained-release forms of these drugs. However, the sustained-release form of benzodiazepine had a lower risk [69].When a study on senior citizens was further carried out, it confirmed that one out of six individuals had problems associated with their stomach, like constipation, and some had dizziness or nausea when receiving opioid treatment [70].

The overall study suggests that opioids should only be given when all other painkillers have become dysfunctional and the person has severe pain that is unbearable. The usage should be avoided mostly in senior citizens because of the risk of other ailments like heart problems, pneumonia, bone fractures, and bones becoming quite fragile, in some cases requiring hospitalization or even death [71-77]. Its effect, however, can be reduced by using it intermittently and avoiding using sustained-release as much as possible [16].

4. Environmental Impacts

Since pharmaceuticals are used because of their biological activity, they reach the environment because they are often excreted unchanged. The health risks of these compounds as a source of environmental contaminants are poorly understood. Awareness of the presence of pharmaceuticals in the global environment suggests the need for precautionary management action to minimize the release of these contaminants into the environment. The detection of waste pharmaceuticals in the environment creates risk for humans, aquatic life, and wildlife and has now become a serious problem for the pharmacy industry and regulators. There is an urgent need to consider the possible growing effect of several drugs affecting the same receptors [78]. The combination of management strategies along with cultural aspects to enhance the rational usage of drugs would prove to be effective in mitigating the risks presented by pharmaceuticals [79].

Conclusion

On the basis of above discussion, we draw the conclusion that acetaminophen is the frontline analgesic, or rather, the best option available for all types of pain. Since analgesics are commonly and inappropriately used in people of different age groups, they should be used with extreme caution in elderly people. Usage of NSAIDs should be rational and limited to the short-term with proper prescription due to associated side effects. Usually, for severe or moderate pain, opioids (in combination with acetaminophen) may be preferable in patients without substance abuse, but use of non-opioid pain relievers should be encouraged. Studies on opioids as chronic pain relievers should be taken seriously as a potential risk to one's life. However, opioids cannot be spared in the treatment of pain in the case of cancer. Therefore, there is an urgent need to understand the actual need for consumption, and evaluate the rational usage of drugs for a safe life.

References

[1] Holloway, K. A. (2011). "Combating inappropriate use of medicines." *Expert Rev Clin Pharmacol,*4, 3:335–48. https://doi.org/10.1586/ecp.11.14.

[2] Kim, S. D., Cho, J., Kim. I. S., Vanderford, B. J., & Snyder, S. A. (2007). "Occurrence and removal of pharmaceuticals and endocrine disruptors in South Korean surface, drinking and waste waters." *Water research,* 41, 5: 1013-1021.DOI: 10.1016/j.watres.2006.06.034.

[3] Buhrer, C., Endesfelder, S., Scheure, T., & Schmitz, T. (2021). "Paracetamol (Acetaminophen) and the Developing Brain." *International Journal of Molecular Science*, 22 : 11156. https://doi.org/10.3390/ijms222011156.

[4] Wu, S., Zhang, L., & Chen, J. (2012). "Paracetamol in the environment and its degradation by Micro-organisms." *Applied Microbiology and Biotechnology,* 96, 4: 875-884. DOI: 10.1007/s00253-012-4414-4.

[5] Chandrashekharan, N. V., Dai, H., Roos, K. L., Roos, K. L. T., Evanson, N. K., Tomsik, J., Elton, T. S., & Simmons, D. L. (2002). "COX-3, a cyclooxygenase-1 variant inhibited by acetaminophen and other analgesic/antipyretic drugs: Cloning, structure, and expression." *Biological Sciences,* 99, 21: 13926-13931. https://doi.org/10.1073/pnas.162468699.

[6] American Geriatrics Society Panel on Pharmacological Management of Persistent Pain in Older Persons. (2009). "Pharmacological management of persistent pain in older persons." *J Am Geriatr Soc.,* 57:1331–1346. doi: 10.1111/j.1532-5415.2009. 02376.x.

[7] Jordan, K.M., Arden, N.K., & Doherty, M. (2003). "EULAR recommendation 2003: an evidence based approach to the management of knee osteoarthritis: Report

of a Task Force of the Standing Committee for International Clinical Studies Including Therapeutic Trials (ESCISIT)." *Ann Rheum Dis.*,62:1145–1155. doi: 10.1136/ard.2003.011742.

[8] Zhang, W., Doherty, M., & Arden, N. (2005). "EULAR evidence based recommendations for the management of hip osteoarthritis: report of a task force of the EULAR Standing Committee for International Clinical Studies Including Therapeutics (ESCISIT)." *Ann Rheum Dis.*, 64:669–681. doi: 10.1136/ard.2004.028886.

[9] National Institute for Health and Clinical Excellence (NICE). (2008). "Osteoarthritis: The care and management of osteoarthritis in adults." *NICE clinical guideline 59.* London: NICE http://www.nice.org.uk/nicemedia/pdf/CG59NICE guideline.pdf.

[10] American Society of Orthopedic Surgeons. *Treatment of osteoarthritis of the knee (non-arthroplasty).* https://www.aaos.org/research/guidelines/GuidelineOAKnee.asp.

[11] Zhang, W., Moskowitz, R.W., & Nuki, G. (2007)."OARSI recommendation for the management of hip and knee osteoarthritis, Part I: critical appraisal of existing treatment guidelines and systematic review of current research evidence." *Osteoarthritis Cartilage,*15:981–1000. doi: 10.1016/j.joca.2007.06.014.

[12] Zhang, W., Nuki, G. & Moskowitz, R.W. (2010)."OARSI recommendations for the management of hip and knee osteoarthritis, Part III: changes in evidence following systematic cumulative update of research published through January 2009." *Osteoarthritis Cartilage.*18:476–499. doi: 10.1016/j.joca.2010.01.013.

[13] Smith, H.S. (2009). "Potential analgesic mechanisms of acetaminophen." *Pain Physician.*12: 269–280. Potential analgesic mechanisms of acetaminophen - PubMed (nih.gov).

[14] Mitchell, S.J., Hilmer, S.N., Murnion, B.P., & Matthews, S. (2011). "Hepatotoxicity of therapeutic short-course paracetamol in hospital inpatients: impact of aging and frailty." *J Clin Pharm Ther.*, 36:327–335. doi:10.1111/j.1365-2710.2010.01193.x.

[15] Larson, A. M., Polson, J., & Fontana, R. J. (2005). "Acetaminophen-induced acute liver failure: results of a United States multicenter, prospective study." *Hepatology, 42*:1364–1372. DOI:10.1002/hep.20948.

[16] Nobuko, O., & Tatsuro, K. (2020). "Analgesic effect of Acetaminophen: A review of known mechanism and action." *Front Pharmacol.*, 20, 3. https://doi.org/10.3389/fphar.2020.580289.

[17] Algren, D. Adam (2020). "Acetaminophen toxicity." In *Toxicology Cases for the Clinical and Forensic Laboratory*, Edited by Ketha, H., & Grag, U, 75-77. Stacy Masucci. DOI: 10.1016/B978-0-12-815846-3.00035-1.

[18] Li, C., & Martin, B.C. (2011). "Trends in emergency department visits attributable to acetaminophen overdoses in the United States: 1993–2007." *Pharmacoepidemiol Drug Saf.*, 20:810–818. DOI:10.1002/pds.2103.

[19] Budnitz, D.S., Lovegrove, M.C., & Crosby, A.E. (2011). "Emergency department visits for overdoses of acetaminophen-containing products." *Am J Prev Med.*, 40:585–592. DOI:10.1016/j.amepre.2011.02.026.

[20] The Acetaminophen Hepatotoxicity Working Group. *Recommendations for FDA Interventions to Decrease the Occurrence of Acetaminophen Hepatotoxicity*. http://www.fda.gov/downloads/AdvisoryCommittees/CommitteesMeetingMaterials/Drugs/DrugSafetyandRiskManagementAdvisoryCommittee/ucm161518.pdf.

[21] Larson, A.M. (2007). "Acetaminophen hepatotoxicity." *Clin Liver Dis.,*11:525–548. DOI:10.1016/j.cld.2007.06.006.

[22] Briant, R. H., Dorrington, R. E., Cleal, J., & Williams, F. M. (1976). "The rate of acetaminophen metabolism in the elderly and the young." *J Am Geriatr Soc.,*24:359–362. https://scholar.google.com/scholar_lookup?journal.

[23] Divoll, M., Ameer, B., Abernethy, D. R., & Greenblatt, D. J. (1982). "Age does not alter acetaminophen absorption." *J Am Geriatr Soc.,*30:240–244. doi: 10.1111/j.1532-5415.1982.tb07093.x.

[24] Wynne, H.A., Cope, L.H., Herd, B., Rawlins, M.D., James, O.F., & Woodhouse, K. W. (1990). "The association of age and frailty with paracetamol conjugation in man." *Age Ageing,* 19:419–424. doi: 10.1093/AGEING/19.6.419.

[25] Miners, J. O., Penhall, R., Robson, R. A., & Birkett, D. J. (1988). "Comparison of paracetamol metabolism in young adults and elderly males." *Eur J Clin Pharmacol,* 35:157–160. DOI:10.1007/BF00609245.

[26] Bannwarth, B., Pehourcq, F., & Lagrange, F. (2001)."Single and multiple dose pharmacokinetics of acetaminophen (paracetamol) in polymedicated very old patients with rheumatic pain." *J. Rheumatol,* 28:182–184.Single and multiple dose pharmacokinetics of acetaminophen (paracetamol) in polymedicated very old patients with rheumatic pain - PubMed (nih.gov).

[27] Liukas, A., Kuusniemi, K., & Aantaa, R. (2011). "Pharmacokinetics of intravenous paracetamol in elderly patients." *Clin Pharmacokinet,* 50:121–129. DOI:10.2165/11537240-000000000-00000.

[28] Divoll, M., Greenblatt, D. J., Ameer, B., & Abernethy, D. R. (1982). "Effect of food on acetaminophen absorption in young and elderly subjects." *J Clin Pharmacol,* 22:571–576. DOI:10.1002/j.1552-4604.1982.tb02651.x.

[29] McEvoy, L., Carr, D. F., & Pirmohamed, M. (2021). "Pharmacogenomics of NSAID-Induced Upper Gastrointestinal Toxicity." *Front. Pharmacol.*, https://doi.org/10.3389/fphar.2021.684162.

[30] Daniel, H., & Solomon, M. D. (2017). "The Risk of Major NSAID Toxicity with Celecoxib, Ibuprofen, or Naproxen: A Secondary Analysis of the PRECISION Trial." *The American journal of medicine*, 12: 1415-1422. https://doi.org/10.1016/j.amjmed.2017.06.028.

[31] Marcum, Z. A., Perera, S., & Donohue, J. M. (2011). "Analgesic use for knee and hip osteoarthritis in community-dwelling elders." *Pain Med.,*12:1628–1636. DOI:10.1111/j.1526-4637.2011.01249.x.

[32] Marcum, Z. A., & Hanlon, J. T. (2010). "Recognizing the risks of chronic non-steroidal anti-inflammatory drug use in older adults." *Ann Long Term Care,*18:24–27103.28.Recognizing the Risks of Chronic Nonsteroidal Anti-Inflammatory Drug Use in Older Adults - PubMed (nih.gov).

[33] Grosser, T. (2006). "Non-steroidal anti-inflammatory." In *Applied Pharmacokinetics and Pharmacodynamics: Principles of therapeutic drug monitoring.* Edited

by Michael E. Burton. Baltimore: Lippincott Williams & Wilkins. https://www.worldcat.org/title/applied-pharmacokinetics-pharmacodynamics-principles-of-therapeutic-drug-monitoring/oclc/59148565.

[34] Woodhouse, K. W., & Wynne, H. (1987). "The pharmacokinetics of non-steroidal anti-inflammatory drugs in the elderly." *Clin Pharmacokinet.*,12:111–122. DOI:10.2165/00003088-198712020-00002.

[35] Rahme, E., Barkun, A., Nedjar, H., Gaugris, S., & Watson, D. (2008). "Hospitalizations for upper and lower GI events associated with traditional NSAIDs and acetaminophen among the elderly in Quebec, Canada." *Am J Gastroenterol,* 103:872–882. DOI:10.1111/j.1572-0241.2008.01811.x.

[36] Rahme, E., & Nedjar, H. (2007). "Risks and benefits of COX-2 inhibitors vs non-selective NSAIDs: does their cardiovascular risk exceed their gastrointestinal benefit? A retrospective cohort study." *Rheumatology (Oxford,)* 46:438. DOI:10.1093/rheumatology/kel428.

[37] Lévesque, L. E., Brophy, J. M., & Zhang, B. 2005. "The risk for myocardial infarction with cyclooxygenase-2 inhibitors: a population study of elderly adults." *Ann Intern Med.* 142:481–489. DOI:10.7326/0003-4819-142-7-200504050-00113.

[38] Lévesque, L. E., Brophy, J. M., & Zhang, B. (2006). "Time variations in the risk of myocardial infarction among elderly users of COX-2 inhibitors." *CMAJ.*, 174:1563–1569. DOI:10.1503/cmaj.051679.

[39] Mamdani, M., Jurrlink, D. N., & Lee, D. S. (2004). "Cyclo-oxygenase-2 inhibitors versus non-selective non-steroidal anti-inflammatory drugs and congestive heart failure outcomes in elderly patients: a population-based cohort study." *Lancet.*, 363:1751–1756. DOI:10.1016/S0140-6736(04)16299-5.

[40] Varas-Lorenzo, C., Riera-Guardia, N., & Calingaert, B. (2011). "Stroke risk and NSAIDs: a systematic review of observational studies." *Pharmacoepidemiol Drug Saf.*, 20:1225–1236. DOI:10.1002/pds.2227.

[41] Solomon, D. H., Rassen, J. A., Glynn, R. J., Lee, J., Levin, R., & Schneeweiss, S. (2010). "The comparative safety of analgesics in older adults with arthritis." *Arch Intern Med.*,170:1968–1978. doi: 10.1001/archinternmed.2010.391.

[42] Schneider, V., Lévesque, L. E., Zhang, B., Hutchinson, T., & Brophy, J. M. (2006). "Association of selective and conventional nonsteroidal antiinflammatory drugs with acute renal failure: A population-based, nested case-control analysis." *Am J Epidem.,* 164:881–889. DOI:10.1093/aje/kwj331.

[43] Gooch, K., Culleton, B. F., & Manns, B. J. 2007. "NSAID use and progression of chronic kidney disease." *Am J Med.* 120:280. DOI:10.1016/j.amjmed.2006.02.015.

[44] Kerr, S. J., Rowett, D. S., Sayer, G. P., Whicker, S. D., Saltman, D. C., & Mant, A. (2011). "All-cause mortality of elderly Australian veterans using COX-2 selective or non-selective NSAIDs: a longitudinal study." *Br J Clin Pharmacol,* 71:936–942. DOI:10.1111/j.1365-2125.2010.03702.x.

[45] Hanlon, J. T., Schmader, K. E., & Landerman, L. R. (1997). "Relation of prescription nonsteroidal anti-inflammatory drug use to cognitive function among community-dwelling elderly." *Ann Epidemiol.,* 7:87–94. DOI:10.1016/S1047-2797(96)00124-X.

[46] Hanlon, J. T., Backonja, M., Weiner, D., & Argoff, C. (2009). "Evolving pharmacological management of persistent pain in older persons." *Pain Med.*,10:959–961. DOI: 10.1111/j.1526-4637.2009.00698.x.

[47] Pergolizzi, J., Böger, R. H., & Budd, K. (2008). "Opioids and the management of chronic severe pain in the elderly: consensus statement of an International Expert Panel with focus on the six clinically most often used World Health Organization Step III opioids (buprenorphine, fentanyl, hydromorphone, methadone, morphine, oxycodone)." *Pain Pract.*, 8:287–313.doi: 10.1111/j.1533-2500.2008.00204.x.

[48] Reid, M. C., Bennett, D. A., Chen, W. G., Basil, A., & Eldadah, M. D. (2011). "Improving the pharmacologic management of pain in older adults: identifying the research gaps and methods to address them." *Pain Med.*,12:1336–135.doi: 10.1111/j.1526-4637.2011.01211.x.

[49] Hilmer, S. N., Gnjidic, D., & Abernethy, D. R. (2010). "Pharmacoepidemiology in the postmarketing assessment of the safety and efficacy of drugs in older adults." *J Gerontol A Biol Sci Med Sci.*, 67:181–188. DOI:10.1093/gerona/glr066.

[50] Benyamin, R.,Trescot, A. M., Datta, S., Buenaventura, R., Adlaka, A., Sehgal, N., Glaser, S. E., & VallejoR. 2008. "Opioid complications and side effects." *Pain physician* 11, 2:S105-20. https://pubmed.ncbi.nlm.nih.gov/18443635/#article-details.

[51] McLachlan, A. J., Bath, S., & Naganathan, V. (2011). "Clinical pharmacology of analgesic medicines in older people: impact of frailty and cognitive impairment." *Br J Clin Pharmacol,* 71:351–364. DOI:10.1111/j.1365-2125.2010.03847.x.

[52] Saari, T. I., Ihmsen, H., Neuvonen, P. J., Olkkola, K. T., & Schwilden, H. (2012). "Oxycodone clearance is markedly reduced with advancing age: a population pharmacokinetic study." *Br J Anaesth.*, 108:491–498. DOI:10.1093/bja/aer395.

[53] Liukas, A., Kuusniemi, K., & Aantaa, R. (2008). "Plasma concentrations of oral oxycodone are greatly increased in the elderly." *Clin Pharmacol Ther.*, 84:462–467. DOI:10.1038/clpt.2008.64.

[54] Liukas, A., Kuusniemi, K., & Aantaa, R. (2011). "Elimination of intravenous oxycodone in the elderly: a pharmacokinetic study in postoperative orthopedic patients of different age groups." *Drugs Aging,* 28:41–50. DOI: 10.2165/11586140-000000000-00000.

[55] Scott, J. C., & Stanski, D. R. (1987). "Decreased fentanyl and alfentanil dose requirements with age. A simultaneous pharmacokinetic and pharmacodynamic evaluation." *J Pharmacol Exp Ther.*, 240:159–165. https://pubmed.ncbi.nlm.nih.gov/3100765/.

[56] Likar, R., Wittels, M., Molnar, M., Kager, I., Ziervogel, G., & Sittl, R. (2006)." Pharmacokinetic and pharmacodynamic properties of tramadol IR and SR in elderly patients: a prospective, age-group-controlled study." *Clin Ther.*, 28:2022–2039. DOI:10.1016/j.clinthera.2006.12.007.

[57] Crews, K. R., Gaedigk, A., & Dunnenberger, H. M. (2011). "Clinical Pharmacogenetics Implementation Consortium (CPIC) guidelines for codeine therapy in the context of cytochrome P450 2D6 (CYP2D6) genotype." *Clin Pharmacol Ther.*, 91:321–326. DOI:10.1038/clpt.2011.287.

[58] American Geriatrics Society. (2012) *Updated Beers Criteria For Potentially Inappropriate Medication Use in Older Adults.*

[59] Hanlon, J. T., & Weiner, D. K. (2009). "Methadone for chronic pain in older adults: blast from the past but are we ready for it to return to prime time?" *Pain Med.,*10:287–288. DOI:10.1111/j.1526-4637.2009.00560.x.

[60] Clegg, A., & Young, J. B. (2011). "Which medications to avoid in people at risk of delirium: a systematic review." *Age Aging,*40:23–29. DOI:10.1093/ageing/afq140.

[61] Morrison, R. S., Magaziner, J., & Gilbert, M. (2003). "Relationship between pain and opioid analgesics on the development of delirium following hip fracture." *J Gerontol A Biol Sci Med Sci.,*58:76–81. DOI: 10.1093/gerona/58.1.M76.

[62] Takkouche, B., Montes-Martínez, A., Gill, S. S., & Etminan, M. (2007). "Psychotropic medications and the risk of fracture: a meta-analysis." *Drug Saf.,* 30:171–184. DOI:10.2165/00002018-200730020-00006.

[63] Miller, M., Stürmer, T., Azrael, D., Levin, R., & Solomon, D. H. (2011). "Opioid analgesics and the risk of fractures in older adults with arthritis." *J Am Geriatr Soc.,*59:430–438. DOI:10.1111/j.1532-5415.2011.03318.x.

[64] Saunders, K. W., Dunn, K. M., & Merrill, J. O. (2010). "Relationship of opioid use and dosage levels to fractures in older chronic pain patients." *J Gen Intern Med.,*25:310–315. DOI:10.1007/s11606-009-1218-z.

[65] Solomon, D. H., Rassen, J. A., & Glynn, R. J. (2010). "The comparative safety of opioids for nonmalignant pain in older adults." *Arch Intern Med.,* 170:1979–1986. DOI:10.1001/archinternmed.2010.450.

[66] Buckeridge, D., Huang, A., & Hanley, J. (2010). "Risk of injury associated with opioid use in older adults." *J Am Geriatr Soc.,*58:1664–1670. DOI:10.1111/j.1532-5415.2010.03015.x.

[67] Hanlon, J. T., Boudreau, R. M., & Roumani, Y. F. (2009). "Number and dosage of central nervous system medications on recurrent falls in community elders: the Health, Aging and Body Composition study." *J Gerontol Med Sci.,*64:492–498. DOI: 10.1093/gerona/gln043.

[68] Dublin, S., Walker, R. L., Jackson, M. L., Nelson, J. C., Weiss, N. S., Kroff, M. V., & Jackson, L. A. (2011)."Use of opioids or benzodiazepines and risk of pneumonia in older adults: a population-based case-control study." *J Am Geriatr Soc.,*59:1899–1907. DOI:10.1111/j.1532-5415.2011.03586.x.

[69] Papaleontiou, M., Henderson, C. R., & Turner, B. J. (2010). "Outcomes associated with opioid use in the treatment of chronic noncancer pain in older adults: a systematic review and meta-analysis." *J Am Geriatr Soc.,*58:1353–1369. DOI: 10.1111/j.1532-5415.2010.02920.x.

[70] Dosa, D. M., Dore, D. D., Mor, V., & Teno, J. M. (2009). "Frequency of long-acting opioid analgesic initiation in opioid-naïve nursing home residents." *J Pain Symptom Manage.,*38:515–521. DOI:10.1016/j.jpainsymman.2008.11.008.

[71] Nadeau, S. E., Wu, J. K., & Lawhern, R. A. (2021). "Opioids, & Chronic Pain: An Analytic Review of the Clinical Evidence, Front." *Pain Res.,* https://doi.org/10.3389/fpain.2021.721357.

[72] Pengetal, Z. (2018). "Patient-Controlled Intravenous Analgesia for Advanced Cancer Patients with Pain: A Retrospective Series Study." *Pain Research and Management*, 2018. 10.1155/2018/7323581.

[73] Rennick, A., Atkinson, T., Cimino, N. M., Strassels, S. A., McPherson, M. L., & Fudin, J. (2016). "Variability in opioid equivalence calculations," *Pain Medicine,* 17,5:892–898. doi: 10.1111/pme.12920.

[74] Sousa, A. M., Neto, J. D. S., Guimaraes, G. M., Cascudo, G. M., Neto, J. O., & Ashmawi, H. A. (2014). "Safety profile of intravenous patient-controlled analgesia for breakthrough pain in cancer patients: a case series study." *Supportive Care in Cancer.*, 22, 3: 795–801. https://limk.springer.com/article/10.1007/s00520-013-2036-8.

[75] Brummett, C. M., Waljee, J. F., & Goesling, J. (2017). "New persistent opioid use after minor and major surgical procedures in US adults." *JAMA Surg.,* 152:e170504. doi: 10.1001/jamasurg.2017.0504.

[76] Lawal, O. D., Gold, J., Murthy, A., Ruchi, R., Bavry, E., & Hume, A. L. (2020). "Rate and risk factors associated with prolonged opioid use after surgery. A systematic review and meta-analysis." *JAMA Netw Open.* 3:e207367. doi: 10.1001/jamanetworkopen.2020.7367.

[77] Winkelman, T. N. A, Chang, V. W., & Binswanger, I. A. (2018). "Health, polysubstance use, and criminal justice involvement among adults with varying levels of opioid use." *JAMA Network Open.* 1:e180558. doi: 10.1001/jamanet workopen.2018.0558.

[78] Doughton, C. G., & Ternes, T. A. (1999). "Pharmaceuticals and personal care products in the environment; Agents of subtle change." *Environmental Health Perspectives,* 107: 907-938. http://dx.doi.org/10.2307/3434573.

[79] Kemmerer, K. (2009). "Pharmaceuticals from human use in the environment-present knowledge and future challenges." *J. Environ. Manag.*, 90, 8 : 2354-2366. DOI: 10.1016/j.jenvman.2009.01.023.

Chapter 6

A Study of Diclofenac Economy in the Light of Its Indirect Harmful Effects on Vultures, Fishes, and Human Beings

Pratima Ghosh[1,*] and Manjula Upadhyay[2]

[1]Department of Commerce, Navyug Kanya Mahavidyalaya, Lucknow, India
[2]Navyug Kanya Mahavidyalaya, Lucknow, India

Abstract

Diclofenac, or DCF, is a widely used non-steroidal anti-inflammatory drug (NSAID) that is also used in the treatment of cattle and for veterinary purposes. It was invented in the U.S.A and was first sold as a generic medicine in 1988. The sudden decrease and mass extinction of vultures in the late 20th century led scientists to find out the reason for their extinction. It was found to be diclofenac. The present chapter attempts to analyze the harmful effects of diclofenac on vultures as well as on fishes and human beings. The study also tries to find out the contribution of the drug to the revenue generation of the economy. The methodology used is Marxian dialectics and the data source is secondary.

The findings suggest that the vultures fall prey to diclofenac by consuming the carcasses of the animals treated with DCF. They suffer from renal failure within 4 days. The same is the case with fish in the water. The findings also suggest that governments all over the world have taken action in the form of banning the use of DCF in cattle treatment. Substitutes for DCF have been invented. All that is required is proper implementation. Yet it is true that the DCF industry in India, along with its substitutes, is growing fast due to an increase in demand, especially by humans.

* Corresponding Author's Email: pratimaghosh2003@gmail.com.

In: Pharmaceuticals: Boon or Bane
Editor: Neha Agarwal
ISBN: 979-8-88697-487-4

Keywords: diclofenac, veterinary, vultures, cattle treatment, economy

1. Introduction

Pharmaceutically active compounds (PhACs) are extensively and increasingly used in human and veterinary medicine [1] and represent a class of emerging environmental contaminants [2]. PhACs can enter the environment by different routes. They can ultimately end up in drinking water, groundwater, and even soil. In the present context, the environmental fate of DCF in different compartments such as soil and water, human beings and birds has been addressed with an overview of current treatment methods. In addition, the toxicity concerns regarding DCF in aquatic as well as terrestrial environments have been analyzed. DCF is a widely used NSAID that is also used in the treatment of cattle and for other veterinary purposes. Diclofenac sodium is a benzene-acetic acid derivative (Novartis). This sodium is a white or slightly yellowish crystalline powder with low solubility in water at 25°C. Its chemical name is 2-[(2, 6-dichlorophenyl) amino] benzene acetic acid, monosodium salt (molecular weight 318.14) and has the molecular formula $C_{14}H_{10}Cl_2NaO_2$. It has the following structural formula, as mentioned in Figure 1.

The inactive ingredients in Diclofenac sodium delayed-release tablets include hydroxypropyl methylcellulose, lactose monohydrate, magnesium stearate, methacrylic acid copolymer, microcrystalline cellulose, polyethylene glycol, povidone, propylene glycol, sodium starch glycolate, talc, titanium dioxide, and triethyl citrate [3].

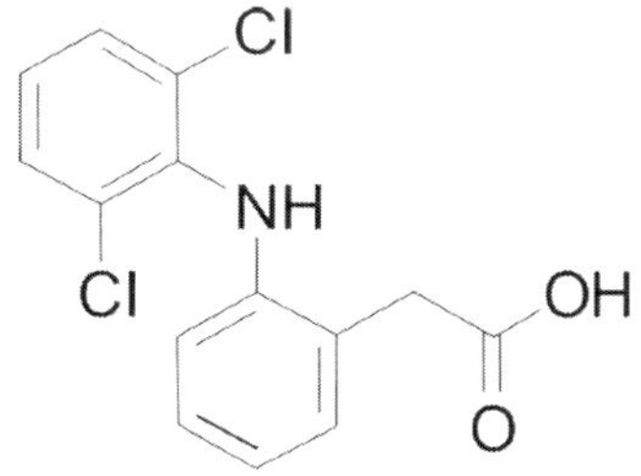

Figure 1. Diclofenac.

2. Objective of Study

The study seeks to examine how far the sustainable development goals of life on land and life on water have been achieved in India, with special reference to diclofenac medicine. That is, the medicine has been found to be harming the fish and vultures through renal failure. India and its neighboring countries like Nepal and Bangladesh have witnessed this problem. The study tries to find out the impact of diclofenac on human beings. The study also examines the impact of diclofenac on the business and economic environment.

2.1. Hypothesis

H0: Diclofenac has been dangerous to aquatic and terrestrial animals.
H1: Sodium is not dangerous to aquatic and terrestrial animals.
H0: Diclofenac contributes a substantial amount to the revenue generated by the pharmaceutical industry.
H1: The contribution of diclofenac in pharmacy has been reduced.
H0: There is no alternative to diclofenac.
H1: There is an alternative to diclofenac.

2.2. Methodology

The methodology of the study is Marxian dielectrics (Britannica 1911), which distinguishes between the method of inquiry and the method of presentation. The method of enquiry is an extensive study of reference materials available. Earlier, the author was unable to access literary works due to the COVID-19 pandemic, but the internet has served its purpose. Along with some restrictions, some open-access journals and some websites, such as Google Scholar and Academia Research Gate, provide free access to the authors.

2.3. The Environmental Concerns

DCF is a commonly used NSAID used for the therapy of chronic forms of arthritis and mild-to-moderate and acute pain. Therapy with DCF in full doses is frequently associated with mild serum aminotransferase elevations and, in rare instances, can lead to serious clinically apparent, acute or chronic liver disease [4, 5, and 6]. India's DCF industry is known for two reasons; one good and one bad. First, the industry is thriving because it provides sodium medicines to the elderly, which act as painkillers; second, the country's growing elderly population and an increase in the number of operations in the country have increased the demand for DCF. On the other hand, increased use of DCF in the treatment of animals has resulted in the killing of vultures because when these animals die, vultures act as scavengers. The main question is what should be the optimal use of DCF in a country like India. Various studies have proven DCF to be very ubiquitous in various environmental concerns. There has been intensive research in the past few decades. It has become an emerging environmental contaminant because of its frequent occurrence in freshwater environments. It has been hazardous for fish, mussels, and various other organisms. Chronic exposure to DCF has been linked to severe effects, despite the fact that at typical detected environmental concentrations, the drug has no toxic effects. About 30% to 70% removal has been obtained through conventional treatment systems in the plants related to wastewater treatment, and they have been the primary sink. The sodium reacts with other contaminants to form new contaminants in the wastewater treatment plant, leading to possible emerging contaminants. Various classes of PhACs such as beta-receptor blockers, antibiotics, psycho-active drugs, lipid regulators, analgesics, NSAIDs, and hormone regulators are frequently detected in raw sewage, treated wastewater, drinking water, and rivers throughout the world with concentrations ranging from a few ng/L to μg/L [7].

According to the study of south Indian River water, Yerabham and Praveen Kumar Reddy (2021) explain that, in most cases, concentrations of pharmaceutical compounds in outlets were found to be less than in inlets [8]. That may be due to the aeration due to the activated sludge process. In a few cases, the NSAID concentrations were higher in the effluents than in the clarifier. Birgit Hoerger et al. [9] studied fish that need to be exposed to the same mechanism of action as in the case of mammals. If they are exposed to DCF, they can be endangered with kidney failure, as the study has seen in the aquatic environment of Germany in trout. The Logical Indian (2018) [10] claims the global intake of DCF at its peak was 2,400 tonnes annually. Only a small amount of human waste is filtered out by treatment facilities, and more than a hundred tonnes of human waste remains in water bodies. As per the study, 20% is absorbed by marine life and natural ecosystems, and the remaining waste flows toward the oceans. In the treatment of wastewater concentrations of effluence, human metabolism has been found. The effects of a membrane bioreactor (MBR) on conventional activated sludge solid retention time (SRT) and hydraulic retention time (HRT) have been analyzed and their effects suggest that only a small portion is changed into sludge MBR and attached growth bioreactors may result in higher elimination of DCF. Long SRT (>150 d) favours elimination due to sludge adaptation. Longer HRTs (>2-3 d) could significantly increase the elimination of DCF during biological wastewater treatment [11].

3. Terrestrial Effects of Diclofenac

It has been found that in India and Nepal since 2000, many species of vultures have died. Tests have been undertaken in nine states and three species: *Gyps bengalensis, Gyps indicus,* and *Gyps tenuirostris* [12]. A simulation model of vulture demography demonstrates that the observed rates of population decline could be caused by contamination with a lethal level of DCF in a small proportion (between 1:130 and 1:760) of ungulate carcasses available to vultures [13]. Visceral gout was found in every dead vulture sample even after the drug was banned. But it has been observed that the intensity of the DCF has been reduced after the ban [14]. "Down to Earth" reports that it is not only DCF but also three other chemicals used for medical purposes that affect the vultures. The report says that after a time period of sixteen years, the problem has again revived and that the vultures were vulnerable to not only DCF but also to other three drugs, namely aceclofenac, ketoprofen, and nimesulide. These drugs were alternatives to diclofenac.

India's oldest veterinary society, the Bombay Natural History Society (BNHS), a biodiversity conservation group, wrote a letter urging the Union Ministry of Environment, Forest and Climate Change (MoEF & CC) to ban the use of three veterinary drugs known to kill vultures in the country. Vultures were quite common till the 1980s. Currently, eight species in the country face extinction, says Rinkita Gurav, manager, raptor conservation, World Wide Fund for Nature, India. The country's vulture population crashed from over 40,000 in 2003 to 18,645 in 2015, as per the last vulture census conducted by intergovernmental body, Bird Life. Diclofenac, which is only permitted for human use, is being misused for cattle treatments. Vibhu Prakash, Deputy Director of Bombay Natural History Society, says that the use of such drugs is prescribed by quacks. Diclofenac continues to threaten the remaining vultures of the region. So, scientists have come up with another alternative. The first vulture-safe alternative to diclofenac, meloxicam, was identified in

2006. This is a vulture-safe drug, which offers hope for India's dwindling population of avian scavengers.

Scientists have been working to identify vulture-safe alternatives to the DCF ever since the discovery, in 2003, that the drug was the principal cause of a catastrophic collapse in vulture populations across South Asia. The birds ingest the drug (which is used to treat pain and inflammation in livestock and humans) while scavenging the carcasses of animals treated with it. Four species of South Asian vulture – *white-rumped (or white-backed), slender-billed, Indian and red-headed* – are now listed as critically endangered by the "International Union for Conservation of Nature" after their numbers crashed in the 1990s. The population of white-rumped vultures in India (estimated to be 4 crore in 1991) fell by 99.9% between 1992 and 2007, the fastest population crash ever recorded for any bird species. After this discovery, in 2006, diclofenac was banned for livestock in India, Pakistan, and Nepal, and in 2010 in Bangladesh. However, enforcement of the ban has been a persistent problem. In addition, other NSAIDs, still in use legally, have been shown to be similarly deadly to vultures [15, 16].

4. Significance of Tolfenamic Acid

4.1. Diclofenac and Human

Sarah Spitz (2013), in her study, has named an increase in the risk of contracting *Clostridium difficile* along with heart attack as the result of DCF in humans [17]. This is highly contagious and resistant to antibiotics. It is considered a superbug that can cause intestinal infections that are found to be severe in about 10 percent of patients and fatal in about 5 percent of cases. This drug is effective in multiple indications. In rheumatology, they are used in the treatment of a range of diseases, including rheumatoid arthritis, lupus erythematosus, psoriatic arthritis, ankylosing spondylitis, gout, and rheumatic fever. The authors of the study concluded that diclofenac at a dose of 150 mg per day is currently the most effective drug in the treatment of pain and physical disability caused by osteoarthritis (OA), and superior to widely used NSAIDs (including ibuprofen, naproxen, and celecoxib) [16]. The findings in the journal Biological Conservation suggested by Prof. Buchely suggest that there is a spurt in feral dogs following the decline of vultures [18]. These dogs normally feed on carcasses that are disease-ridden which has increased their numbers by seven million. Following this, there was a rabies outbreak in India that killed 48,000 people between 1992 and 2006. The deaths may have been avoided if not for the disappearance of vultures. This happened in Sub-Saharan Africa also, but in a much more challenging way. The researchers term it a darker story. The deaths of vultures can allow other scavengers like rats, ants, etc. to flourish, which can bring bacteria and viruses from carcasses into human cities.

There is an increasing spread of anthrax, brucellosis, and foot and mouth diseases due to the vacuum created by vulture extinction. There is a very significant finding that high body temperatures and speedy metabolism with strong stomach acids allow vultures to ingest carrion infected by anthrax bacteria without any ill effects. The metabolism of vultures is a dead-end for pathogens, whereas dogs and insects tend to become carriers of the same. The culture of Parsis is also at stake through this disaster. This Zoroastrian community is finding it difficult to sustain their unique funeral ritual where the dead are disposed of in towers of silence, to be consumed by scavenger birds, mainly vultures. According to Parsis, cremating the dead on land or water pollutes nature.

The finding that NSAID tolfenamic acid is also safe for vultures is the result of a systematic safety testing study led by the Indian Veterinary Research Institute, in collaboration with the Bombay Natural History Society and the Royal Society for the Protection of Birds. Details of the study were published in a preprint paper in August. Prakash V. investigated this and concluded that there is very little hope for their revival.

It is "hugely significant" that tolfenamic acid has been confirmed as vulture-safe, said Saving Asia's Vultures from Extinction, a consortium of 24 organizations working on vulture conservation, in a report on its website. Saving Asia's vultures from extinction's programmed manager and co-chair of the International Union for Conservation of Nature's Vulture Specialist Group, explained to 'The Third Pole', "Vets often complain that they like to choose between drugs for different situations. Each NSAID has slightly different properties" [19]. Deaths caused by NSAIDS are invisible. Birds die two-three days after ingesting the medicine, making it difficult to establish a clear link," says Chris Bowden, co-chair, Vulture Specialist Group at the International Union for Conservation of Nature. Diclofenac has been administered to captive vultures in experiments, which died after feeding on tissues of domestic animals that had received a normal veterinary dose of the drug a few hours before death.

5. Diclofenac Market

According to the forecast (2022-2027), the Indian DCF sodium industry is growing at a steady rate. An increasing elderly population and an increase in the number of surgeries have increased the demand for DCF through the forecasted period of 2026. Sports injuries are also demanding of DCF on a regular basis. Besides this, supporting government policy is also one of the main factors. Improvement in medical facilities has also been a lucrative factor for the increase in the demand for medicine in the fiscal year 2022-2027. A CAGR of 3.6 percent is expected [20].

The following segments have been identified to administer diclofenac in the Indian market.

1. Administrative route identification number
2. Route of indication route
3. Distribution channel
4. Geographical area

Diclofenac is consumed the most by developed nations in North America. The reason may be an increase in the intake of junk food, which harms the teeth and bones, resulting in diclofenac administration. This is followed by Europe, where the consumption is 30%. Asia-Pacific consumes 28 percent of diclofenac, and South America consumes only 5%. The rest of the world consumes only 5% of DCF. This has been explained with the help of the following pi chart mentioned in Figure 2.

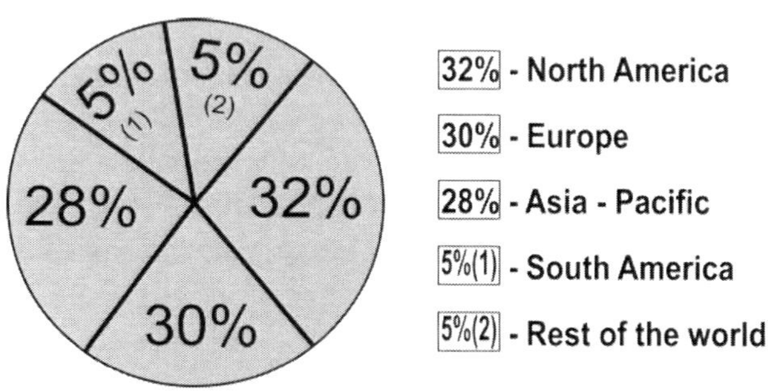

Figure 2. Consumption of DCF (Source, calculated from various reports of market watch).

5.1. DCF Market Size in India

Scitechresearch (2020) has tried to present the details of DCF use in the category of historic years from 2016 to 2019, base year i.e., 2021, expectations for 2021-22, and forecast for 2022 to 2024. According to the report, the growth of DCF will be driven during the forecast period due to growing cases of arthritis. As far as manufacturing is concerned, the contract manufacturing organization segment is expected to witness significant growth on account of outsourcing activities and competitive price offerings. Major companies are getting approval by regulatory bodies like ICMR and launching new products in order to remain competitive. Other strategies include mergers, acquisitions and research collaboration. The major players operating in the Indian DCF sodium market are Novartis India Ltd, Hetero Drugs Ltd., Auro Laboratories Ltd., Aarti Drugs Ltd., J B Chemicals and Pharmaceuticals Ltd., EmkayPharma Pvt. Ltd., Bestochem Formulations (India) Limited, Salvavidas Pharmaceutical Pvt. Ltd., WellonaPharma Pvt. Ltd. and others. Major companies are working to develop advanced technologies [21, 22].

The pain relief manufacturing and marketing in India was around rupees 4200 crore in 2017-19 and is expected to grow at CGAR of 3.9% till 2024 (Economic times, 26th September 2019). Table 1 shows the result of a covert survey done in Uttar Pradesh (one of the major sates of India in terms of population and fauna) to find out what percentage of pharmacies are selling DCF, Meloxicam, Nimesulide, Piroxicam and a combination of other medicines for cattle treatment.

Table 1. The percentage of pharmacies that offered for sale the four most common NSAIDs in Uttar Pradesh (adopted from Article *in* Bird Conservation International September 2020)

2012-2017(Covert Survey)	2012	2013	2015	2016	2017
N	50	11	89	32	59
Diclofenac	72	18.2	31.5	21.9	20.3
Meloxicam	6	18.2	12.4	3.1	27.1
Nimesulide	6	9.1	29.2	3.1	23.7
Piroxicam	14	54.5	20.2	71.9	25.4
Other	2		6.7		3.4

N = the number of pharmacies.

'Other' NSAIDs include: aceclofenac, metamizole (Analgin), ketoprofen, mefenamic acid, phenylbutazone and tolfenamic acid.

Table 2. Segments of Indian diclofenac market

Prescription	By Form	ByApplication	By Route of Administration	By Source	By Distribution Channel	ByEnd User	By Region
OTC	Tablet	Osteoarthritis	Oral	In-house	Online	Adult	North
	Injection	Ankylosing Spondylitis	Intravenous	Contract Manufacturing Organizations	Offline	Paediatric	South
	Gel	Dysmenorrhea	Topical				East
	Others	Rheumatoid arthritis					West
		Migraine					
		Pain					
		Others					

It has been found that there has been awareness, the number of diclofenac pharmacies have declined from 71 percent to 27 percent, and the use of meloxicam has increased from 6 to 27 percent. The same is the case with other drugs as well. This shows the positive response to the ban being imposed on DCF.

The Indian DCF sodium market has been segmented into the following categories, in addition to the industry trends, which have also been detailed below in Table 2.

5.2. Competition in the Diclofenac Market

Techsciresearch has performed both primary as well as exhaustive secondary research for this study. Initially, Tech Sci Research sourced a world list of manufacturers. Subsequently, research was conducted with primary research surveys with the identified companies. While interviewing, the respondents were also asked about their competitors. Through this technique, it could include the manufacturers that could not be identified due to the limitations of secondary research. Tech Sci Research analyzed the manufacturers, distribution channels, and presence of all major players across the globe. Bottom Up, wherein data for various end-user segments is recorded and forecasted for the future years. Tech Sci Research sourced these values from industry experts and company representatives and externally validated them by analyzing historical data of these product types and applications to get an appropriate, overall market size. Various secondary sources such as company websites, news articles, press releases, company annual reports, investor presentations, and financial reports were also studied by the organization. [23, 24].

5.3. Findings

The findings of the research say that there has been a steady increase in the Indian market of the drug since 2017, and is forecasted to increase at the same pace and will more than double by 2026.

The hypothesis that the drug under study is dangerous to fishes, vultures, and human beings, is true as it is found in ample evidence.

There is significant scope for DCF and related drugs in revenue generation, so our null hypothesis is true and accepted.

Our third hypothesis is also true that meloxicam and tolfenamic acid are nearly the perfect substitutes for diclofenac with less danger.

Conclusion

Identification of alternative drugs is a must to treat cattle and other animals. There is an importunate need to provoke awareness pertaining to the use of DCF, until this problem is rectified. Captive holding and breeding of vultures are recommended as a precaution to ensure the long-term survival of the threatened species and to provide a stock of birds for future reintroduction programs. Since India manufactures the maximum veterinary NSAIDs that are used in South Asia, a ban on DCF, aceclofenac, or ketoprofen should be considered by the Indian government. Putting a ban on DCF is an integral part of India's National Action Plan for Vulture Conservation (2019–2024) and its role in the Regional Declaration on the Conservation of South Asia's Critically Endangered Vulture Species (Anon 2012), which commits signatories to identify and prevent the veterinary use of all vulture-toxic NSAIDs. Further studies are required to better assess the fate and toxicological effects of DCF and its metabolites, and must consider the possible use of the substitutes. On the other hand, for the treatment of sewage water, bioaugmentation should be used to enhance DCF elimination. There is an urgent need to conduct further research on deconjugation processes, biotic and abiotic transformation products, and on microbial communities that degrade diclofenac in a green manner.

References

[1] Hollender, J. et al. (2010). "H. & LC–high-resolution MS in environmental analysis: from target screening to the identification of unknowns." *Anal Bioanal Chem,* 397: 943–951. https://doi.org/10.1007/s00216- 10-3608-9.

[2] Lonappan, L., Brar, S.K., Das, R.K., Verma, M., & Surampalli, R.Y. (2016). "Diclofenac and its transformation products: Environmental occurrence and toxicity - A review." *Environ Int.*, 96: 127-138. doi: 10.1016/j.envint.2016.09.014.

[3] Google. 2021. "DiclofenacDelayed Release Tablets" Last updated November 22. https://www.drugs.com/pro/diclofenac-delayed-release-tablets.html.

[4] Weigel, S., Kuhlmann, J., Hühnerfuss, H. (2002). "Drugs and personal care products as ubiquitous pollutants: occurrence and distribution of clofibricacid, caffeine, and

DEET in the North Sea." *Sci. Total Environ.*, 295, 1-3: 131–141. doi: 10.1016/S0048-9697(02)00064-5.

[5] Willis, J.V., Kendall, M.J., Flinn, R.M., Thornhill, D.P., Welling, P.G. (1979). "The pharmacokinetics of diclofenac sodium following intravenous and oral administration." *Eur. J. Clin. Pharmacol.*, 16, 6: 405–410. doi: 10.1007/BF00568201.

[6] Xu, J., Wu, L., Chang, A.C. (2009). "Degradation and adsorption of selected pharmaceuticals and personal care products (PPCPs) in agricultural soils." *Chemosphere,* 77, 10: 1299–1305.

[7] Barbosa, L.M.C., Souto, N., Ferreira, R.M.A., Scarpassa, V.M. (2016). "Behavioral patterns, parity rate and natural infection analysis in anopheline species involved in the transmission of malaria in the northeastern Brazilian Amazon region." *Acta Tropica*, 164: 216-225. https://doi.org/10.1016/j.actatropica.2016.09.018.

[8] Reddy, Y.P.K., Vimalkumar, K., Ramaswamy, B.R., Kumar, V., Singhal, R.K., Basu, H., Gopal, C.M., Vandana, K.E., Bhat, K., Udayashankar, H.N., & Balakrishna, K. (2021). "Assessment of non-steroidal anti-inflammatory drugs from selected wastewater treatment plants of Southwestern India." *Emerging Contaminants*, 7: https://doi.org/10.1016/j.emcon.2021.01.001.

[9] Hoeger, B., Köllner, B., Dietrich, D.R., & Hitzfeld, B. (2005). "Water-borne diclofenac affects kidney and gillintegrity and selected and immuneparameters in brown trout (Salmotruttafario)." *Aquat Toxicol,* 75, 1: 53-64. doi: 10.1016/j.aquatox.2005.07.006. PMID.

[10] Kukreti, S.K. (2018). "Pharma Waste Threatening Rivers Globally, Saysstudy." *The LogicalIndian*. https://thelogicalindian.com/environment/pharma-waste-threatening%20-rivers-globally.

[11] Vieno, N., & Sillanpää, M. (2014). "Fate of diclofenac in municipal wastewater treatment plant-A review." *Environment International*, 69: 28–39. https://doi.org/10.1016/j.envint.2014.03.021.

[12] Cuthbert, R. J., Taggart, M. A., Saini, M., Sharma, A., Das, A., Kulkarni, M. D., Deori, P., Ranade, S., Shringarpure, S.N., Galligan, T.H., & Green, R.E. (2016). "Continuing mortality of vultures in India associated with illegal veterinary use of diclofenac and a potential threat from nimesulide." *Oryx, 50,1*: 104–112. doi: 10.1017/S003060531500037X.

[13] Cuthbert, R.J., Dave, R., Chakraborty, S.S., Kumar, S., Prakash, S., Ranade, S. P. & Prakash, V. (2011). "Assessing the ongoing threat from veterinary non-steroidal anti inflammatory drugs to critically endangered Gyps vultures in India." *Oryx*, 45:420–426. doi: 10.1017/S0030605311000135.

[14] Gurav, R. (2021). "Drug debacle: Endangeredvulture population stillunderthreat." *Wildlife & Biodiversity*. https://www.downtoearth.org.in/blog/wildlife-biodiversity/drug-debacle-endangered-vulture-population-still-under-threat-78835.

[15] Triebskorn, R., Casper, H., Heyd, A., Eikemper, R., Kohler, H.R., Schwaiger, J. (2004). "Toxic effects of the nonsteroidal anti-inflammatory drug diclofenac. Part II: cytological effects in liver, kidney, gills and intestine of rainbow trout (*Oncorhynchus mykiss*)." *Aquat. Toxicol.*, 68, 2: 151–166. doi: 10.1016/j.aquatox.2004.03.015.

[16] Valdés, M.E., Amé, M.V., Bistoni, M.d.l.A., Wunderlin, D.A. (2014). Occurrence and bioaccumulation of pharmaceuticals in fish species inhibiting the suquia river basin (Córdoba, Argentina)." *Sci. Total Environ.*, 472: 389–396. doi: 10.1016/j.scitotenv.2013.10.124. Epub 2013 Dec 1.

[17] Spitz, S. (2013). "Diclofenac linked to increased risk of C. difficile." *CMAJ*, 185, 6: 471. doi: 10.1503/cmaj.109-4426.

[18] Bowden, C., & Botha, A. (2022). "Vulture Specialist Group IUCN Species Survival Commission." https://www.researchgate.net/publication/360167393_Vulture_Specialist_Group_IUCN_Species_Survival_Commission.

[19] Special Correspondent. (2014). "Uttar Pradesh bans veterinary use of Diclofenac. *The Hindu*, September 29. https://www.thehindu.com/news/national/uttar-pradesh-bans-veterinary-use-of-diclofenac/article6455231.ece.

[20] The report: "Diclofenac Market Forecast (2022-2027)," by Industry ARC coversan in-depth analysis of the following segments of the Diclofenac Market. https://www.drugs.com/pro/diclofenac.html.e-of-diclofenac/article6455231.ece.

[21] Google. 2022."DiclofenacMarketForecast (2022-2027)." *Industry ARC*, Report code- HC 75845. https://www.drugs.com/pro/diclofenac.html.

[22] Spitz, S. (2013). "Physicians should be cautious when prescribing diclofenac." *CMAJ.*, 185, 6: 470. doi: 10.1503/cmaj.109-4425. Epub 2013 Mar 4.

[23] Galligan, Toby H., John W. Mallord, Vibhum. Prakash, Krishna P. Bhusal, A. B. M. Sarowar Alam, Fergus M. Anthony, Ruchi Dave, Alka Dube, Kartik Shastri, YogeshKumar, Nikita Prakash, Sachin Ranade, Rohan Shringar pure, Devendra Chapagain, Ishwary P. Chaudhary, Ankitb. Joshi, Khadananda Paudel, Tarik Kabir, Sakibahmed, Kazi Z. Azmiri, Richard J. Cuthbert, Christopher G. R. Bowden and Rhys E. Green.(2021). "Trends in the availability of the vulture-toxic drug, diclofenac, and other NSAIDs in South Asia, as revealed by covert pharmacy surveys." *Bird Conservation International, 31,* 3:337-353. https://doi.org/10.1017/S0959270920000477.

[24] India Diclofenac Sodium Market By Product Type (Prescription v/s OTC), By Form (Tablet, Injection, Gel, Others), By Application (Osteoarthritis, Ankylosing Spondylitis, Dysmenorrhea, Rheumatoid arthritis, Migraine, Pain and Others), By Route of Administration (Oral, Intravenous, Topical, Others), By Source (In-house v/s Contract Manufacturing Organizations), By Distribution Channel (Online v/s Offline), By End User (Adult v/s Paediatric), By Company, By Region, Forecast & Opportunities, FY2026. https://www.techsciresearch.com/report/india-diclofenac-sodium-market/7197.html.

Chapter 7

Cocrystals of Pharmaceutical Drugs: A Case Study of Paracetamol Cocrystals with Oxalic Acid and 4, 4, Bipyridine

Keshav Kumar Singh*, **Preeti Prajapati and Poonam Tandon**
Department of Physics, University of Lucknow, Lucknow, India

Abstract

This chapter aims to describe cocrystallization as a method to improve the physical and chemical properties of pharmaceutical drugs to make them more effective. Cocrytsllization alters the crystal structure of the drug, and therefore, this method has the potential to modify the physicochemical properties of drugs to make them more bioavailable and reactive. This chapter briefly summarizes "what constitutes an effective drug" in the first section. The second section describes the idea of cocrystallization and its advantages. In this review, cocrystals of paracetamol with oxalic acid and 4,4 bipyridine are described in detail, demonstrating how paracetamol's reactivity and tableting properties improve when it is cocrystallized with oxalic acid and 4,4 bipyridine. This study also demonstrates the need for further research into pharmaceutical cocrystals, which can greatly improve the existing properties of medicines.

Keywords: pharmaceutical cocrystals, density functional theory, Paracetamol, crystal engineering, mechanical properties

---

* Corresponding Author's Email: keshavsinghi007@gmail.com.

In: Pharmaceuticals: Boon or Bane
Editor: Neha Agarwal
ISBN: 979-8-88697-487-4

1. Introduction

A substance used for the diagnosis, treatment, or prevention of diseases is known as a pharmaceutical drug [1]. Drugs are primarily organic compounds, generally divided into broad classes of small organic molecules and the molecules derived from biological systems such as insulin glargine, etc. Inorganic and organometallic compounds are also sometimes used as drugs, for which primary examples include lithium and platinum-based agents such as lithium carbonate and cisplatin, as well as gallium.

However, the study of medicinal chemistry is largely focused on the small organic system. Drug discovery was a revolutionary step that has not only been used to cure or prevent diseases but also increased the life span of mankind. Advanced medicinal facilities and therapies have given a boon to human lives. Over the past decades, pharmaceuticals have emerged as the most reliable companion for humans across the globe. Pharmaceutical sciences combine various scientific disciplines to discover new drugs and therapies. They are synthetic or maybe natural chemicals that have therapeutic properties and are developed as medications to be administered to patients to cure them and alleviate their illnesses. Pharmaceutical drugs are classified into various categories based on pharmacological usage [2]. These are kept in different groups according to their chemical structures and how they are used to cure a specific disease. Some of them are listed in Table 1.

2. Salient Properties of Drugs

Drug properties comprise the structural, physicochemical, biochemical, pharmacokinetic (PK), and toxicity characteristics. In addition, drug properties such as solubility, permeability, metabolic stability, and transporter effects are important parameters for the success of drug candidates. These, in turn, affect oral bioavailability, metabolism, clearance, toxicity, as well as *in vitro* pharmacology [3].

2.1. Solubility

Solubility is one of the most challenging properties in drug discovery for the formulations of poorly soluble drugs. Many approaches have been discovered, such as size reduction, solid dispersion, complexation, salt formation, nanoparticles, self-emulsifying drug delivery systems (SEDDS), co-solvents, nano-suspension, and emulsion and cocrystal formation for the solubility enhancement of poorly water-soluble drugs. Each technique has its merits and demerits, and certain factors should be kept in mind for selecting the technique [4].

2.2. Stability

A stability study is extremely important during the development of a new dosage formulation. Several stability studies should be performed during the development of a drug, such as relative humidity stress, chemical stability, thermal stability, solution stability, and photo stability study. For example, to determine the effect of water on the formulation of a drug, automated water sorption and desorption studies are performed under relative humidity stress [5].

Table 1. Classification of pharmaceutical drugs

Type	Pharmacological Action
Antibiotic	Used to treat or prevent bacterial infections
Antimicrobial	Used to kill microorganisms or stop their growth
Anti-inflammatory	Used for the treatment of inflammation or swelling
Antipyretic	Used to reduce fever
Antidepressant	Used to treat major depressive disorders
Anticancer	Used to destroy and inhibit cancer cells
Antiviral	Used for treating viral infections
Antipsychotic	Used to manage psychosis (abnormal conditions of mind)
Analgesics	Used to provide pain relief
Antidiabetic	Used to treat diabetic mellitus by altering glucose levels
Anticholesteremic	Used to lower lipid levels to reduce cardiovascular events
Antifungal	Used to cure fungal infections
Antiasthmatic	Used to treat or prevent asthma attacks
CNS drugs	Used to regulate the central nervous system
Tranquilizers	Used to treat anxiety and tension

2.3. Bioavailability

Bioavailability is the rate and extent of a pure drug that reaches the specific therapeutic site of action. The low oral bioavailability of drug substances is the primary challenge during the development of new solid forms [6]. Almost 60-70% of the drugs have poor aqueous solubility and low oral bioavailability. Most of them fall into BCS (biopharmaceutical classification system), class II (high permeability/low solubility) and class IV (low permeability/low solubility).

2.4. Permeability

Permeability deals with drug absorption and distribution across the biological membrane. Compounds with poor permeability tend to have low oral bioavailability, low oral absorption, less cell membrane penetration, and less exposure to the target area [7].

2.5. Melting Point

The melting point is a physical property of a drug. It demonstrates thermodynamic stability and is used to determine the purity of drug materials [8]. The high melting point of drugs determines their thermodynamic stability, i.e., thermal stability, and a low melting point reflects less thermal stability.

2.6. Tabletability

Crystal packing, tabletability, and compaction are important parameters during preformulation studies [9]. In addition, the mechanical properties of drug substances are often determined by their tabletting properties.

2.7. Metabolism

Metabolism affects clearance and oral bioavailability of compounds. Rapid clearance and low oral bioavailability can limit the exposure of drugs to the target area and they become less effective [10]. On the other hand, extremely low metabolism can cause prolonged half-life and drug accumulation in the body that results in toxicity.

3. Cocryslallization: A Method to Improve Physiochemical Properties of Drugs

Poor aqueous solubility and low oral bioavailability of an active pharmaceutical ingredient (API) are the major obstacles during the development and designing of new solid forms. So far, several methods for modifying the physical and chemical properties of APIs have been discovered, including size reduction, solid dispersion, complexation, salt formation, nanoparticles, self-emulsifying drug delivery system (SEDDS), addition of co-solvents, nano-suspension, emulsion, and cocrystal formation. Among all these approaches, cocrystallization is a unique pathway that offers a great opportunity to develop new drugs with superior physicochemical properties such as solubility, bioavailability, stability, melting point, tabletability, etc.

3.1. What are API and Coformer?

API is the term that refers to the biologically active component of a pharmaceutical drug [11]. Drug products are comprised of several components. The API is the key ingredient, and the other component is the excipient, which is usually pharmacologically inactive and safe, such as lactose and mineral oil.

A coformer is an additional component which is added to an active pharmaceutical ingredient in a stoichiometric ratio via some non-covalent interactions to modulate the physicochemical properties of APIs. Therefore, selecting a coformer is important for designing and screening new solid forms. These should be non-toxic with no adverse side effects. In addition, It must be approved and generally regarded as safe (GRAS) [12].

3.2. What is Pharmaceutical Cocrystal?

The term "cocrystal" was first discovered by Etter to designate the rule of hydrogen bonding in organic compounds [13]. Later, G. Desiraju gave the supramolecular synthon concept of hydrogen bond formation in crystal structures [14]. Pharmaceutical cocrystals are described as a distinct class of novel crystalline materials which could alter the API's physicochemical properties. Pharmaceutical cocrystals are a composition of multiple molecular complexes in which an API and a coformer are bound together in the crystal lattice through non-covalent interactions [15-17]. The cocrystal components are present in a definite stoichiometric ratio and are discrete as well as neutral molecular reactants which remain solid at ambient temperature. These have different intermolecular spacing patterns and, therefore, exhibit different physical and chemical properties than pure components.

3.3. Advantages of Cocrystals

The various approaches developed by scientists and researchers to enhance the physicochemical properties of APIs are salt formation, micronization, solid dispersion, amorphous drugs, and encapsulation. Among all these, cocrystallization is a unique process that offers many advantages, such as:

(i) Cocrystals will exist in a stable crystalline form with no need for other excipients and additives in formulations.
(ii) They modulate the physicochemical properties such as melting point, tabletability, solubility, stability, bioavailability, and permeability of APIs without altering their pharmacological properties. The APIs will benefit from their physicochemical property enhancements because of the presence of coformers in their crystal structures, which is a property modifying component.
(iii) Cocrystals can be made for non-ionisable APIs as well as for those drugs that have sensitive functional groups that may not survive the harsh reaction conditions of strong acids or bases.
(iv) They have the potential to shorten the drug development timeline.
(v) They are cost-effective, which is appealing to pharmaceutical companies.
(vi) They can be classified as green chemistry as they offer high yield, no solvent use, and few by-products.

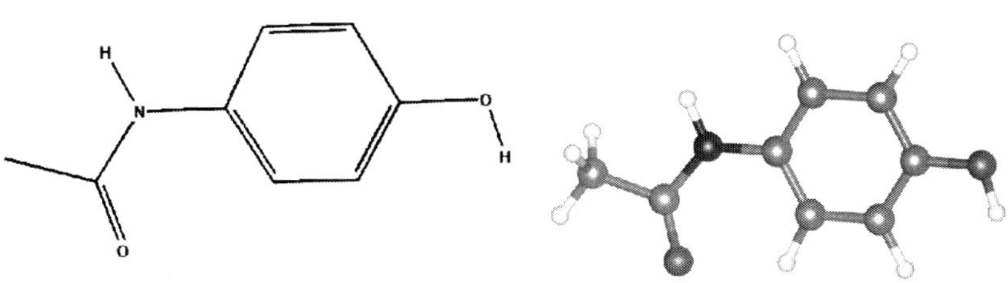

Figure 1. 2D & 3D structures of paracetamol.

In the present chapter, an example of paracetamol is discussed as a case study showing how cocrystallization affects the drug properties.

Table 2. Common physical properties of paracetamol [22]

Chemical formula	$C_8H_9NO_2$
Appearance	White odourless crystalline powder; large monoclinic prisms from water
Molecular weight	151.16
Melting point	169-170.5°C
Density	1.293gm/cc
Solubility	Soluble in water (1:70, 1:20 at 100°C), ethanol (1:7), acetone (1:13), chloroform (1:50), glycerol (1:40), methanol (1:10), propylene glycol (1:9) and solutions of alkali hydroxides; insoluble in diethylether, slightly soluble in ether. It is insoluble in petroleum ethers, pentone, and benzene.
Dissociation constant pK_a	9.0–9.5

4. Paracetamol

Paracetamol, also known as acetaminophen is a non-steroidal anti-inflammatory drugs (NSAID) which is widely used to treat fever, mild pain, and prevent blood clots [18, 19, 20]. Paracetamol is one of the WHOs list of essential medicines [21]. It has the IUPAC designation of n-(4-hydroxyphenyl) acetamide. The 2d and 3d structures of paracetamol are shown in Figure 1. The common physical properties of paracetamol are tabulated in Table 2.

There are three known polymorphs of paracetamol, of which only two are the stable ones. They are referred as form I and form II of paracetamol, respectively. Their crystal structure is shown in Figure 2.

The crystal structure of form 1 is monoclinic. It has pleated sheets stacked along the b axis, while form 2 of paracetamol has an orthorhombic structure in which hydrogen-bonded flat sheets are stacked along the c axis. Being the most stable, these two polymorphs of paracetamol are of particular industrial importance. By the observation of Figure 2, it can be seen that it is quite difficult to compress the crystal of form 1 paracetamol since the pleated sheets in form 1 make it relatively stiff. Therefore, to make a tablet of form 1, binders such as polyvinyl pyrrolidone (PVP), corn starch paste, and gelatine solution (acacia mucilage) are required. Besides, in the case of form 1, its morphology strongly depends on the saturation level of solution and solvent. Therefore, the manufacturing of the drug itself is solvent and environment-dependent.

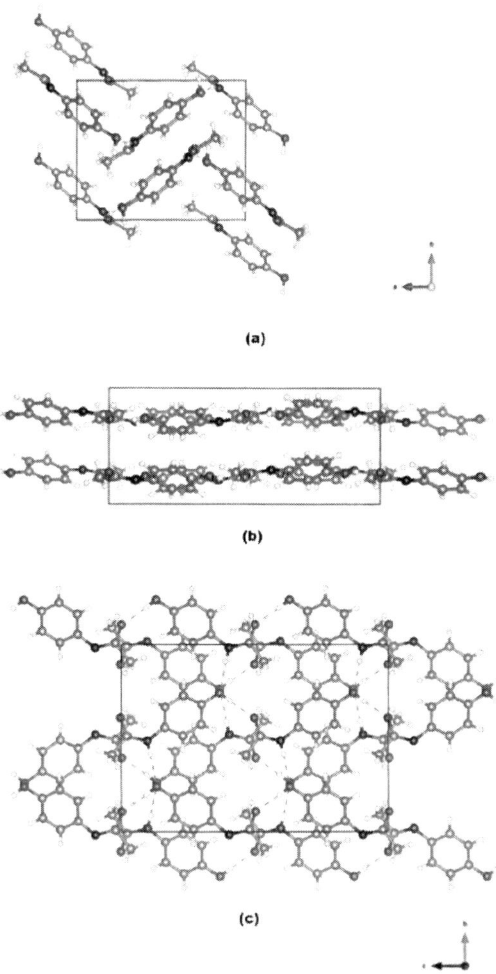

Figure 2. Crystal structure of paracetamol (PRA) form 1 (a), side view of form 2 (b) top view (c).

As it is an organic system, the vibrational spectra of paracetamol serves as the most important characteristic used to identify its presence in any chemical system. Therefore, both the IR and Raman spectra of paracetamol are shown in the Figure 3, while their important features are listed in Table 3.

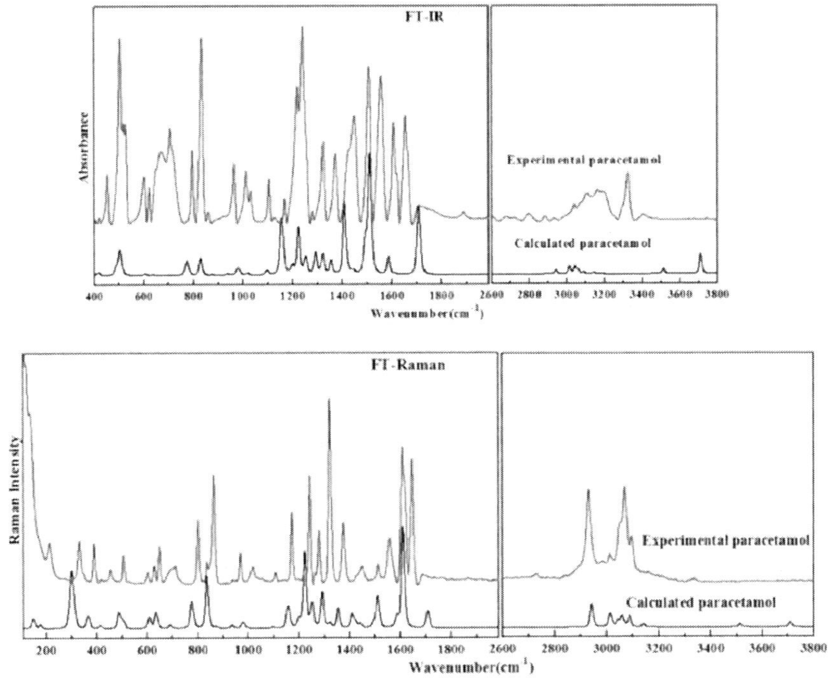

Figure 3. Experimental and calculated IR and Raman Spectra of paracetamol [23].

Table 3. Important IR/RAMAN features of paracetamol [23]

IR/RAMAN features	Experimental frequency	Calculated frequency
OH stretching	3321	3320 (dimer)
NH stretching	3256	3439 (dimer
Asy (CH_3) stretching	3015	3010
Sym (CH_3) stretching	2943	2931
--C=O stretching	1651/1649	1643

Vibrational spectra of paracetamol is also used to study any structural changes that might occur when the structure of paracetamol is altered to improve the characteristics of paracetamol.

4.1. Crystallization in Paracetamol

Crystallization of paracetamol has been used as a method to improve the mechanical and reactivity of the said drug. In the present chapter, two crystallization attempts, namely cocrystal of paracetamol with oxalic acid and 4,4 bipyridine have been discussed. Both of these are used to improve the mechanical properties of form I of paracetamol.

Alpha-oxalic acid (IUPAC designation ethanedioic acid and molecular formula $HO_2C{-}CO_2H$) is an organic acid and the simplest member of the dicarboxylic acid family. It naturally occurs in the wood-sorrel (Oxalis) [24], members of the spinach and the brassica family, and stems of all species of the genus Chenopodium and related genera of the family Amaranthaceae. Many soil fungus species, for example, *Aspergillus niger* also produce oxalic acid [25, 26]. Therefore, oxalic acid becomes one natural choice to use as a conformer since it is already part of many plant's lives, so it should not have an adverse effect on human biology if mixed with a drug. On the other hand, 4,4'-Bipyridine, ($C_{10}H_8N_2$), is a synthetic compound that is primarily used as an industrial precursor to paraquat $[(C_6H_7N)_2]Cl_2$. It is also used in a variety of coordination polymers. It is a colourless solid that can be used as a conformer due to its lack of toxic effect.

Ahmed et al. created the first cocrystal of paracetamol (form 1) with the oxalic acid (PRA-OXA) and 4,4'–bipyridine (PRA-BPY) by preparing a slurry of 30.2 g of PRA and 18 g of OXA in 10ml of ethyl acetate and 30.2 g of PRA and 31.2 g of 4, 4' -bipyridine in the 10 ml of acetonitrile. These mixtures were stirred for 48 h and 18 h respectively, after which the solid was extracted and air dried at room temperature. Various characterizations, including XRD, SEM, and DSC were carried out to determine the purity of the cocrystals prepared as discussed in the ref [27]. Srivastava et al. [23] studied these cocrystals in depth using density functional theory (at B3LYP/6-311++G(d,p) level of theory [28, 29, 30, 31, 32] by employing the supramolecular synthon approach. In this approach, the basic unit of a molecular crystal is formed by intermolecular interactions. In this case, one unit of PRA and one unit of OXA and BPY are used as an API and conformer to make PRA-OXA and PRA-BPY cocrystals, respectively. In this cocrystal, PRA acts as a hydrogen bond donor via two of its functional groups namely amide–NH and phenol–OH groups, while OXA and BPY act as accepters of hydrogen bonds via their own –OH groups (in OXA) and N of the heterocyclic ring (in BPY) respectively. Srivastava et al. [23] studied the structure and interaction in the cocrystal unit of both PRA-OXA and PRA-BPY cocrystals by utilizing natural bond orbit (NBO) analysis [33], quantum theory of atoms in molecules (QTAIM) [34] and vibrational spectroscopy. Additionally, the reactivity of the newly formed cocrystals was also studied using frontier molecular orbits and global and local reactivity descriptors. The dimer of the same systems was also studied by the same method to improve the correlation between experimental and theoretical studies.

4.2. Geometric Structure and Energies

Figure 4 shows the optimized monomer and dimer structures of the proposed PRA-OXA and PRA-BPY supramolecular synthons. The calculations performed by Srivastava et al. [23] show that in the case of both systems (viz. monomer and dimer), the binding energy is about 19 Kcal/mol for PRA-OXA and 21.8 Kcal/mol for the PRA-BPY system, respectively. This large binding energy shows that both cocrystal systems are stable at room temperature. Also, the structure of the dimer system has almost no deviation from the experimentally reported structure as shown in Figure 4.

The experimental IR and RAMAN spectra corresponding to the PRA-OXA and PRA-BPY systems are shown in Figures 5 and 6 [23]. These spectra were compared with the simulated spectra calculated from monomer and dimer to understand the origin of the various vibrational modes. The assignment of the most important vibrational mode is tabulated in Table 4.

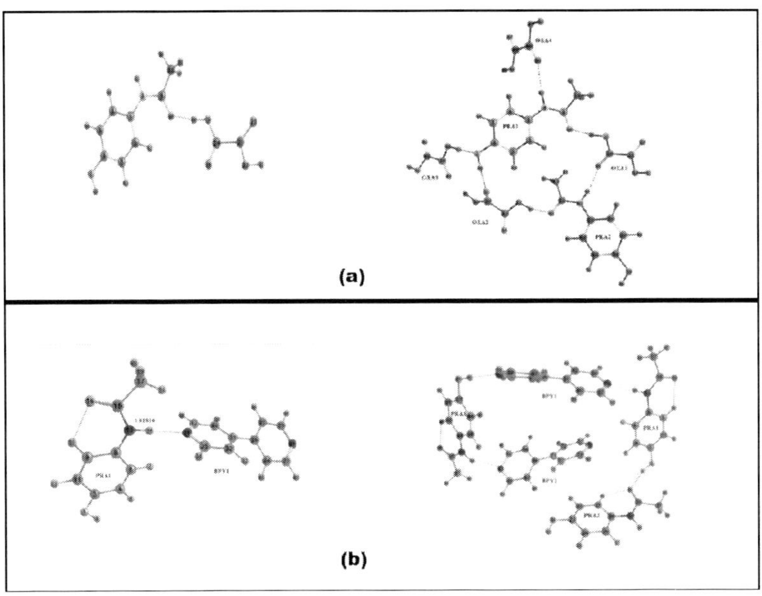

Figure 4. Optimized structure of the monomer (left) and dimer (right) of the PRA-OXA (pallet a) and PRA-BPY (pallet b).

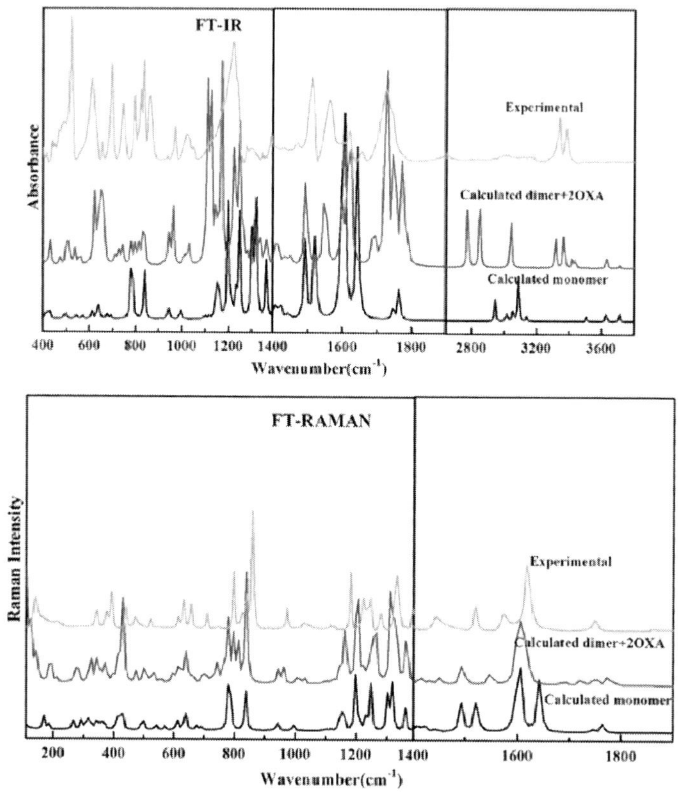

Figure 5. InfraRed (Upper pallet) and Raman (lower pallet) spectra of the monomer and dimer of the PRA-OXA and their comparison with experimental spectra [23].

Only dimer frequencies are reported here as they are closer to the experimental values. The details are given in ref 23. From Table 4, it can be seen the largest shift in the wave number occurs in the OH group of OXA and the NH group of PRA, since these are the bonds involved in the hydrogen bonding interactions responsible for the formation and stability of the cocrystal PRA-OXA. The experimental blue shift in the starching frequencies is 88 and 284 cm^{-1} for the OH group of OXA and the NH group of PRA, while the calculated ones are 183 and 234 cm^{-1}. The blue shift in the vibrational spectra represents the loss of polarizability of the bond involved, which can be due to the charge redistribution in the cocrystal system. In the PRA-BPY system, the aromatic C==N of BPY and the NH group of PRA are responsible for the cocrystal formation. The blue shift in aromatic C==N frequency is about 15 and 20 cm-1 for experimental and calculated values. This represents a weak contribution of aromatic C==N in cocrystallization as no large polarization occurs and electronic states in the aromatic rings of BPY before and after co-crystallization are almost the same. On the other hand, a large redshift (152 cm^{-1} in experimental and 162 cm^{-1} in calculated spectra) was observed in the IR spectra of paracetamol before and after co-crystallization. This large redshift represents a strong hydrogen bond responsible for the PRA-BPY cocrystal stability.

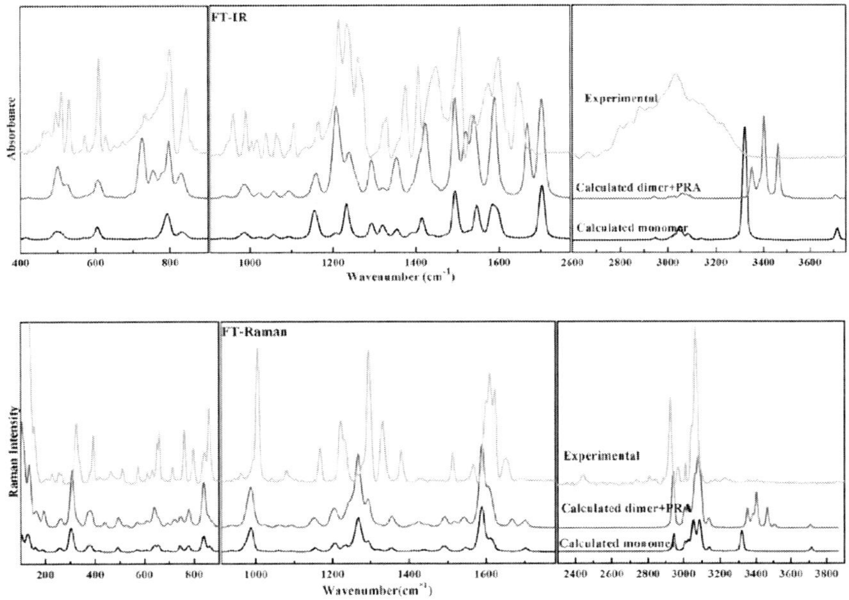

Figure 6. InfraRed (Upper pallet) and Raman (lower pallet) spectra of the monomer and dimer of the PRA-BPYand their comparison with experimental spectra [23].

4.3. Stability and Interactions in Co-Crystal Units

Now that we have complete information about which groups are involved in hydrogen bonding, the next step will be the determination of the strength of these interactions. Natural bond orbit (NBO) analysis [32] was performed by Srivastava et al. [23] on both cocrystal structures in order to study these interactions. NBO analysis is one of the methods for transforming a given set of electronic wave functions into their localized forms. These orbitals are then

Table 4. Most important vibrational (IR/RAMAN) modes

IR/RAMAN Mode						
	PRA-OXA					
	PRA		OXA		Cocrystal	
	Experimental	Calculated	Experimental	Calculated	Experimental	Calculated
OH stretching	3321	3715	3099	3086/3631	3383/3380	3320
NH stretching	3256	3256	…	…	3344/3346	3439
C=O stretching	1690	1651/1649	1749/1739	1790	1721/1773	1616
	PRA-BPY					
	PRA		BPY		Cocrystal	
	Experimental	Calculated	Experimental	Calculated	Experimental	Calculated
OH stretching	3325	3711	…	…	3224	3403
NH stretching	3294	3513	…	…	3142	3351
C=O stretching	1655/1648	1698	…	…	1645/1650	1668
C==N stretching	…	…	1219/1218	1216	1234/1232	1236/1233

Table 5. Inter and intra molecular interactions in the monomer and dimer of PRA-OXA and PRA-BPY structures predicted from NBO analysis

Donor NBO(i)	ED(i)/e(a.u.)	Acceptor NBO (j)	ED(j)/e(a.u.)	Interaction energy (Kcal/mol)	ED(j)-ED(i)(a.u.)
PRA-OXA					
Monomer					
LP(1)O10	1.953	σ *O25-H28	0.087	9.50	1.02
LP(2)O10	1.849	σ *O25-H28	0.087	32.10	0.70
Dimer					
LP(1)O7	1.937	σ *O38-H43	0.066	16.86	0.94
LP(1)O31	1.958	σ *O7-H16	0.046	9.38	1.37
LP(1)O26	1.972	σ *N60-H43	0.024	4.79	1.15
PRA-BPY					
Monomer					
π (N41-C49)	1.723	π*(C46-C47)	0.346	12.58	0.33
LP(1)N13	1.655	π*(C15-O16)	0.301	63.04	0.28
Dimer					
LP(1)O36	1.967	σ*(O1-H2)	0.035	3.64	1.15
LP(2)O36	1.866	σ*(O1-H2)	0.035	8.34	0.74
LP(1)N56	1.908	σ*(N13-H14)	0.030	6.63	0.83
LP(1)N41	1.899	σ*(O81-H82)	0.036	12.04	0.82
LP(1)N61	1.911	σ*(N93-H94)	0.027	4.38	0.81

utilized in the calculation of the distribution of the electron density throughout the chemical system. This will provide information about the bonding pattern in the system in question. The bond energies and other relevant parameters obtained from NBO are tabulated in Table 5. The NBO analysis of the PRA-OXA and PRA-BPY monomer systems reveals one hydrogen bonding interaction in both systems. In the PRA-OXA monomer, interaction is due to the lone pair electron of the oxygen of the C=O group of PRA with the σ* orbital on one of the –OH group of OXA. This interaction has an energy of about 32.1 kcal/mol, representing a very strong hydrogen bond. In the PRA-BPY monomer system, such interaction is caused by the lone pair of N of BPY with π* orbital on the –NH group, resulting in the interaction energy of about 63.3 kcal/mol.Several more interactions are revealed in the case of dimers of both the systems. In the PRA-OXA system, three comparatively weaker interactions were observed, with their interaction energies ranging from 4.79 to 16.86 Kcal/mol (see Table 5). Similarly, for the PRA-BPY dimer, five interactions (listed in table 5) ranging from 3.68 to 12.04 Kcal/mol are observed. Out of these five, LP(2)O36→σ*(O1-H2) with interaction energy 8.34 Kcal/mol and LP(1)N56→σ*(N13-H14) with interaction energy 6.63 Kcal/mol confirms the presence of hydrogen bonds responsible for the formation of co-crystal.

4.4. Reactivity of Co-Crystals and Their Comparison with PRA

Having lots of information about the structure and related interactions, the structure and the interactions resulting in the successful cocrystallization of PRA-OXA and PRA-BPY systems, we turn our attention to the reactivity of the PRA after making its cocrystals. Reactivity analysis is important to understand because cocrystals are made to improve physical and chemical properties. Therefore, the reactivity of the drug should either improve or remains the same. Global reactivity descriptors [33, 34, 35, 36, 37] were calculated for PRA-OXA and PRA-BPY by Srivastava et al. [23]. These global reactivity descriptors were calculated from the frontier molecular orbitals (HOMO and LUMO orbitals) obtained from density functional calculations [38]. The energy of the HOMO in a molecular system is a measure of ionization potential, whereas energy of the LUMO represents electron affinity. Since all chemical interactions/reactions occur due to the transfer of electrons between HOMO and LUMO of interacting molecules, the HOMO-LUMO gap becomes the direct measure of reactivity: less HOMO-LUMO gap represents highly reactive molecules and vice versa. Additionally, global reactivity descriptors calculated from these orbitals provide additional reactivity information.

In the present section, reactivity descriptors of dimers are reported since their structure and vibrational spectra are in good agreement with the experimental spectra. For the PRA-OXA system, the PRA molecule has a HOMO-LUMO gap of 5.23 eV. The HOMO-LUMO gap is reduced to 2.43 eV after the co crystallization with OXA. This significant drop in the HOMO-LUMO gap results in the more reactive PRA when it is a part of a cocrystal system. Similarly, electronegativity (χ) and electrophilicity(ω) increases from 3.30 to 4.59 eV and from 2.08 to 8.45 eV, respectively, from PRA to PRA-OXA cocrystal. This means that cocrystals have an increasing tendency to attract and keep electron pairs, which results in increased reactivity. Their chemical hardness (η) too has decreased from 2.61 to 1.24 eV, representing that the ability of PRA to change electronic state is also decreased.

In the case of the PRA-BPY system, its HOMO-LUMO gap decreases to 3.2 eV and electronegativity (χ) and electrophilicity (ω)increase from 2.30 to 3.69 eV and 2.80 to 5.45 eV, respectively, in the cocrystals compared to the PRA molecule. This represents an increase in the reactivity of the PRA molecule when it is a part of the PRA-BPY cocrystal system.

4.5. Mechanical Properties

One of the primary reasons for synthesizing the paracetamol cocrystals is the poor mechanical properties of form I paracetamol. It is known that a thermodynamically stable PRA tablet cannot be formed by using the form I polymorph due to its stiff nature [39]. Cocrystlization improves the mechanical properties of form I PRA [39, 40]. Ahmed et al. [24] first studied the tableting and compactness behaviour of the PRA-OXA and PRA-BPY cocrystal structures by experimentally measuring tensile strength (eq 1) and tablet porosity (eq 2) via the following equations [41,42]:-

$$\sigma_t = \frac{2F_t}{\pi h_t d_t} \tag{1}$$

$$\varepsilon = 1 - \frac{\rho_{tablet}}{\rho_{app}} \tag{2}$$

In equation 1, F_t is the force needed to crush tablets while h_t and d_t are height and diameter of the tablets respectively. The tablet porosity was obtained from tablet density (ρ_{tablet}) and the apparent powder density (ρ_{app}).

The tabletability of a substance is represented by a plot of tensile strength vs. applied compaction pressure. Similar to the previous experimental observations, Ahmed et al. [24] could not form a tablet from PRA-form I polymorph at any pressure. However, in their study of cocrystals, they observed that the greatest increase in the tensile strength occurs for the PRA-OXA system which is above 2 MPa. The tensile strength for PRA-BPY is close to 1.8 MPa. Therefore, PRA-OXA is comparatively better for tableting. Compactibility, i.e., the ability to form a denser tablet is measured by plotting porosity against ostensible strength. Here again, PRA-OXA has more than 20% porosity and hence has better compactibility at 5MPa compared to the PRA-BPY system, which does not show any change in porosity.

Conclusion

As the human population steadily increases, requirements for better drug formulations are increasing. However, several already-known pharmaceutical drugs have the ability to cure many diseases and can be used to tackle increasing drug demand. However, they are not used because of either poor bioavailability or are difficult to manufacture in the desired form. Co-crystallization in pharmaceutical drugs emerges as a possible solution to both problems. By suitably choosing the conformer and API, the reactivity, as well as physical properties of the drugs, can be altered in the required fashion. In the present review, a brief introduction to pharmaceutical drugs and cocrystals has been presented. A case study is presented for the cocrystals of paracetamol (PRA) with oxalic acid (OXA) and 4,4' bipyridine (BPY). It was observed that cocrystallization of paracetamol improves the reactivity and mechanical properties in both cases (PRA-OXA and PRA-BPY) when compared with the pure form of paracetamol. Out of the two cocrystal structures, PRA-OXA is found to be the most effective since it not only has the highest reactivity but also has better tensile strength and compactibility. These properties make it more suitable for making better tablets in industrial setups and more effective drugs. This example of a form I paracetamol demonstrates how cocrystallization can affect and improve drug properties. Therefore, more research is required in this direction.

References

[1] US Federal Food, Drug, and Cosmetic Act, SEC. 210., (g)(1)(B). (2008).Accessed 17 August 2008.

[2] Medthority.com. (2022). "Database of prescription pharmaceutical products including drug classifications." https://www.medthority.com/drugs. Accessed at 27 July 2022.

[3] Di, L., Edward, Kerns, E. H., and Carter, G. T. (2009). "Drug-like property concepts in pharmaceutical design." *Current pharmaceutical design*, 15, 19, 2184-2194. DOI: 10.2174/138161209788682479.

[4] Savjani, K.T., Gajjar, A. K., and Savjani, J.K. (2012). "Drug solubility: importance and enhancement techniques." *International Scholarly Research Notices.*, article id:195727. https://doi.org/10.5402/2012/195727.

[5] McNamara, D.P., Childs, S. L., Giordano, J., Iarriccio, A., Cassidy, J., Shet, M. S., Mannion, R., O'Donnell, E., and Park, A. (2006). "Use of a glutaric acid cocrystal to improve oral bioavailability of a low solubility API." *Pharmaceutical research*,23, 8, 1888-1897. DOI: 10.1007/s11095-006-9032-3.

[6] George, M., and Abraham, T. E. (2007). "pH sensitive alginate–guar gum hydrogel for the controlled delivery of protein drugs." *International journal of pharmaceutics*, 335, 1-2, 123-129. DOI: 10.1016/j.ijpharm.2006.11.009.

[7] Lipinski, C. A., Lombardo, F., Dominy, B. W., and Feeney, P. J. (2012). "Experimental and computational approaches to estimate solubility and permeability in drug discovery and development settings." *Advanced drug delivery reviews*, 64, 1-3, 4-17. DOI:10.1016/j.addr.2012.09.019.

[8] Abourahma, H., Cocuzza, D. S., Melendez, J., and Urban, J. M. (2011). "Pyrazinamide cocrystals and the search for polymorphs." *Cryst Eng Comm*, 13, 21, 6442-6450. DOI:10.1039/C1CE05598D.

[9] Maeno, Y., Toshiro F., Masatoshi, K., Kentaro, Y., Tatsuaki, T., Tetsuya, O., Toyofumi, S., and Tomono, K. (2014). "Novel pharmaceutical cocrystal consisting of paracetamol and trimethyl glycine, a new promising cocrystal former." *International journal of pharmaceutics*, 473, 1-2, 179-186.DOI: 10.1016/j.ijpharm. 2014.07.008.

[10] Li, D., Trapa, P., Obach, R. S., Atkinson, K., Bi, Y. A., Wolford, A. C., Tan, B., McDonald, T. S., Lai, Y., and Tremaine, L. M. "A novel relay method for determining low-clearance values." *Drug metabolism and disposition*, 40, 9, 1860-1865. DOI: 10.1124/dmd.112.046425.

[11] Chan, E. J., Gao, Q., and Dabros, M. (2014). "Understanding the structure details when drying hydrate crystals of pharmaceuticals–interpretations from diffuse scattering and inter-modulation satellites of a partially dehydrated crystal." *Acta Crystallographica Section B: Structural Science, Crystal Engineering and Materials*, 70, 3, 555-567. https://doi.org/10.1107/S2052520614005125.

[12] Steed, J.W. (2013). "The role of co-crystals in pharmaceutical design." *Trends in pharmacological sciences*, 34, 3, 185-193. DOI: https://doi.org/10.1016/j.tips.2012.12.003.

[13] Mattei, A., and Li, T. (2012). "Polymorph formation and nucleation mechanism of tolfenamic acid in solution: an investigation of pre-nucleation solute association." *Pharmaceutical research*, 29, 460-470. https://doi.org/10.1007/s11095-011-0574-7.

[14] Desiraju, G. R. (1995). "Supramolecular synthons in crystal engineering—a new organic synthesis." *Angewandte Chemie International Edition in English*, 34, 21, 2311-2327. https://doi.org/10.1002/anie.199523111.

[15] Duggirala, N.K., Perry, M. L.,Almarsson, O., and Zaworotko, M. J. (2016). "Pharmaceutical cocrystals: along the path to improved medicines." *Chemical communications*, 52, 4, 640-655. https://doi.org/10.1039/C5CC08216A.

[16] Shemchuk, O., Lixing, S., Tumanov, N., Wouters, J., Braga, D., Grepioni, F., and Leyssens, T. (2020). "Chiral resolution of RS-oxiracetam upon cocrystallization with pharmaceutically acceptable inorganic salts." *Crystal Growth & Design*, 20, 4, 2602-2607. https://doi.org/10.1021/acs.cgd.9b01725.

[17] Qiao, N., Li, M., Schlindwein, W., Malek, N., Davies, A., and Trappitt, G. (2011). "Pharmaceutical cocrystals: an overview." *International journal of pharmaceutics*, 419, 1-2, 1-11. DOI: 10.1016/j.ijpharm.2011.07.037.

[18] Warwick, C. (2008). "Paracetamol and fever management." *The Journal of the Royal Society for the Promotion of health*, 128: 320-323. https://doi.org/10.1177/1466424008092794.

[19] Saragiotto, B.T., Shaheed, C. A., and Maher, C. G. (2019). "Paracetamol for pain in adults." *BMJ*, 367, 16693. DOI: 10.1136/bmj.l6693.

[20] Meremikwu, M. M., and Oyolta, A. (2002). "Paracetamol versus placebo or physical methods for treating fever in children." *Cochrane Database of Systematic Reviews*,2, CD003676. doi: 10.1002/14651858.CD003676.

[21] World Health Organization. (2019). "World Health Organization model list of essential medicines: 21st list 2019. Geneva: WHO. Hdl:10665/325771. WHO/MVP/EMP/IAU/2019.06. Licence: CC BY-NC-SA 3.0 IGO. https://www.who.int/groups/expert-committee-on-selection-and-use-of-essential-medicines/essential-medicines-lists.

[22] Paracetamol (Acetaminophen) - Pharmaceutical Drugs - NCBI Bookshelf (nih.gov) accessed at 27 Jul. 22. https://www.ncbi.nlm.nih.gov/books/NBK526213/.

[23] Srivastava, K., Khan, E., Shimpi, M. R., Tandon, P., Sinha, K., and Velaga, S. P. (2018)."Molecular structure and hydrogen bond interactions of a paracetamol–4, 4′-bipyridine cocrystal studied using a vibrational spectroscopic and quantum chemical approach." *CrystEngComm.*,20, 2, 213-222. https://doi.org/10.1039/C7CE01505D.

[24] Rombauer, I. S., Becker, M. R., Becker, J., and Scott, M. (2000). "Joy of Cooking: All about Chicken." in *Joy of Cooking*: Simon and Schuster. https://www.amazon.com/Joy-Cooking-Fully-Revised-Updated/dp/1501169718.

[25] Dutton, M.V., and Evans, C. S. (1996). "Oxalate production by fungi: its role in pathogenicity and ecology in the soil environment." *Canadian journal of microbiology*, 42, 9, 881-895. DOI: 10.1139/m96-114.

[26] Gadd, G. M. (1999). "Fungal production of citric and oxalic acid: importance in metal speciation, physiology and biogeochemical processes." *Advances in microbial physiology*, 41,47-92. https://doi.org/10.1016/S0065-2911(08)60165-4.

[27] Ahmed, H., Shimpi, M. R., and Velaga, S. P. (2017). "Relationship between mechanical properties and crystal structure in cocrystals and salt of paracetamol." *Drug development and industrial pharmacy*,43, 1, 89-97. https://doi.org/10.1080/03639045.2016.1220568.

[28] Petersson, G. A., Bennett, A., Tensfeldt, Mohammad, A., AlLaham., Shirley, W. A., and Mantzaris, J. (1988). "A complete basis set model chemistry. I. The total energies of closed-shell atoms and hydrides of the firstrow elements." *The Journal of chemical physics*, 89, 4, 2193-2218. https://doi.org/10.1063/1.455064.

[29] Dayantis, J., and Palierne, J. F. (1991). "Monte Carlo precise determination of the end-to-end distribution function of self avoiding walks on the simple-cubic lattice." *The Journal of chemical physics*, 95, 8, 6088-6099. https://doi.org/10.1063/1.461578.

[30] Lee, C., Yang, W., and Parr, R. G. (1988). "Development of the Colle-Salvetti correlation-energy formula into a functional of the electron density." *Physical review B.*,37, 2, 785. https://journals.aps.org/prb/abstract/10.1103/PhysRevB.37.785.

[31] Axel, B. (1996). "Density functional thermochemistry. IV. A new dynamical correlation functional and implications for exact exchange mixing." *The Journal of chemical physics*, 104, 3, 1040-1046. DOI: 10.1063/1.470829.

[32] Foster, J. P., and Weinhold, F. (1980). "Natural hybrid orbitals." *Journal of the American Chemical Society*, 102,24, 7211-7218. https://doi.org/10.1021/ja00544a007.

[33] Pearson, R.G. (1989). "Absolute electronegativity and hardness: applications to organic chemistry." *The Journal of Organic Chemistry*, 54, 6, 1423-1430. https://doi.org/10.1021/jo00267a034.

[34] Geerlings, P., Proft, F. D., and Langenaeker, W. (2003). "Conceptual density functional theory." *Chemical reviews*, 103, 5, 1793-1874. https://doi.org/10.1021/cr990029p.

[35] Parr, R.G., Szentpály, L. V., and Liu, S. (1999). "Electrophilicity index." *Journal of the American Chemical Society*,121, 9, 1922-1924. https://doi.org/10.1021/ja983494x.

[36] Chattaraj, P. K., and Giri, S. (2007). "Stability, reactivity, and aromaticity of compounds of a multivalent superatom." *The Journal of Physical Chemistry A*, 111,43, 11116-11121. DOI: 10.1021/jp0760758.

[37] Padmanabhan, J., Parthasarathi, R., Subramanian, V., and Chattaraj, P. K. (2007). "Electrophilicity-based charge transfer descriptor." *The Journal of Physical Chemistry A*, 111, 7, 1358-1361. https://doi.org/10.1021/jp0649549.

[38] Padmaja, L., Amalanathan, M., Ravikumar, C., and Joe, I. H. (2009). "NBO analysis and vibrational spectra of 2, 6-bis (p-methyl benzylidenecyclohexanone) using density functional theory." *Spectrochimica Acta Part A: Molecular and Biomolecular Spectroscopy*, 74,2, 349-356.DOI: 10.1016/j.saa.2009.06.007.

[39] Di Martino, P., Guyot-Hermann,A. M., Conflant, P., Drache, M., and Guyot, J.C. (1996). "A new pure paracetamol for direct compression: the orthorhombic form." *International journal of pharmaceutics*, 128, 1-2, 1-8. https://jglobal.jst.go.jp/en/detail?JGLOBAL_ID=200902103846676690.

[40] Rasenack, N., and Müller, B. W. (2002). "Crystal habit and tableting behavior." *International journal of pharmaceutics*, 244, 1-2, 45-57. https://citeseerx.ist.psu.edu/viewdoc/download?doi=10.1.1.603.2141&rep=rep1&type=pdf.

[41] Fell, J. T., and Newton, J. M. (1970). "Determination of tablet strength by the diametral-compression test." *Journal of pharmaceutical sciences*, 59, 5, 688-691. https://doi.org/10.1002/jps.2600590523.

[42] Sun, C. C. (2004). "A novel method for deriving true density of pharmaceutical solids including hydrates and water-containing powders." *Journal of pharmaceutical sciences*, 93, 3,646-653.DOI: 10.1002/jps.10595.

Chapter 8

Adverse Effects of Pharmaceutical Industry Waste and Effluent on Ecology

Rashmi Gupta[1,*] and Rajendra Gupta[2]
[1]Department of Botany, BSNV P.G. College, University of Lucknow, Lucknow,
[2]ICAR-Indian Institute of Sugarcane Research, Lucknow, India

Abstract

Pharmaceuticals' use has significantly increased to combat diseases and ill-health across the world. However, figures from around the world show that more than half of all pharmaceuticals prescribed, administered, or sold are disposed of incorrectly. When pharmaceuticals are used improperly or misused, it can have a substantial negative impact on ecosystems and species, especially when unused medications are improperly disposed of. Drug manufacturing industries are one of the important sources of environmental pollution.Pharmaceuticals cannot be removed from wastewater using a single technology.Drug manufacturing chemicals cannot be removed from wastewater treatment facilities; as a result, these chemicals seep into aquatic systems. The main pathway of chemicals into the aquatic environment is through the disposal of effluents from wastewater treatment plants. Pharmaceutical industrial wastewater contains significantly higher levels of pollutants because of the presence of organic matter, which is non-biodegradable. Several studies and data analysis have identified specific chemical components that are toxic to living species. The presence of pharmaceutical effluents in environment is alarming due to their highly toxic nature, which includes genotoxicity and mutagenic potential. Severe pollution occurs when wastewater that has not been properly treated is released into an

* Corresponding Author's Email: rashmibsnv@gmail.com.

In: Pharmaceuticals: Boon or Bane
Editor: Neha Agarwal
ISBN: 979-8-88697-487-4

aquatic environment. These effluents pollute the environment in a way that harms human health, economic growth, and aquatic ecosystems. As more and more treated wastewater is used to irrigate crops for sustainable water management, resistant bacteria might find their way into our food. Complex pharmacological mixes may change behaviour, accumulate in tissues, cause reproductive harm, and impede cell growth when exposed to them over an extended period of time at low concentrations. Research suggests that human reproductive systems may get affected when exposed to pharmaceuticals and other xenobiotics in potable water. These effluents exhibited various chromosomal aberrations, namely c-mitosis, chromosome loss, chromosome break, micro-nucleated cells, etc., indicating the poisonous and genotoxic effects of pollutants present in the pharmaceutical waste water. Thus, we can say that pharmaceutical industries are producing lifesaving drugs on one hand but also producing life threatening wastes and effluents on the other hand.

Keywords: pharmaceutical, pollution, environment, ecology, wastewater, aquatic life

1. Introduction

Pharmaceuticals are usually water-soluble, charged molecules, and many of them passes through treatment plants designed to get rid of traditional pollutants. Pharmaceutical residues can withstand degradation, are persistent, highly soluble in water, and become intricate when mixed with other compounds. Pharmaceutical industries are releasing approximately half of the world's wastewater without recommended pre-treatment. Untreated or poorly treated wastewater when released into the aquatic environment causes severe pollution that adversely affects aquatic life, the ecosystem and public health. Pharmaceutical industrial wastewater contains a high level of pollutants because of presence of non-biodegradable organic matter like antibiotics, other prescription and non-prescription drugs, animal and plant steroids, reproductive hormones, betalactamides, anti-inflammatory, analgesics, lipid regulators, anti-depressants, personal care products, detergent metabolites, flame retardants, and other chemicals, i.e., spent solvents, reaction residues, used filter media, heavy metals (such as mercury, lead, nickel, cadmium, and chromium) etc. The presence of pharmaceutical chemicals in environment is hazardous because of their acute toxicity, including genotoxicity and mutagenic potential. Compounds of pharmaceutical importance are generally produced in batch processes resulting in the existence of a good number of products in wastewaters which are produced in numerous operations, wherein more water is used for cleaning or removal of solid cake, or washing of machines. No single technology is available which can eradicate pharmaceutical effluents from wastewaters. Several studies have found that chemicals present in pharmaceutical effluents have toxic effects on global environment and living organisms. Different concentrations of pharmaceuticals in effluent exhibited various chromosomal aberrations like chromosome break, chromosome loss, micro-nucleated cells etc., indicating the toxic and genotoxic effects of chemicals present within the pharmaceutical effluent. Only a fraction of drug consumed is absorbed; the remainder is usually excreted in unmetabolized form, where it enters raw sewage. Three important pathways through which drugs effluents enter the environment are presented in Figure 1.

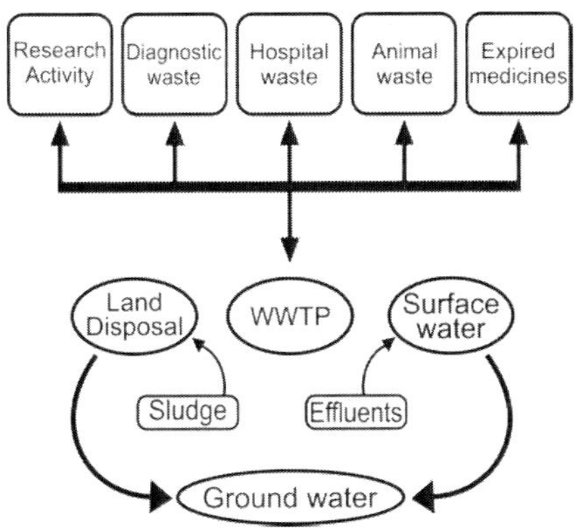

Figure 1. Source of pharmaceutical residues in the environment.
Adapted from Tiwari B. et al. [1].

The largest contributor to water pollution is personal use of pharmaceuticals, where they can enter the water stream through the human body. The second is inappropriate disposal of medicine, like throwing leftover drugs down the sink. The third source is from pharmaceutical industries, which discharge the effluent into water bodies. Approximately 88% of pharmaceuticals in the environment are attributed to personal use and excretion of drugs; 10% are from unused medicines which are not disposed properly, and 2% are attributed to waste generated from pharmaceutical industries. Cardoso et al., [2] reported that surface and ground waters, effluents, sediments, and biota are contaminated by human and veterinary active pharmaceutical ingredients (APIs). More than 2,000 active pharmaceutical ingredients are in use worldwide as prescribed medicines, over-the-counter therapeutic drugs, and veterinary drugs [3]. It is estimated that about 10% of pharmaceutical products have a potential environmental risk [4]. It is also important to know that how long these compounds will persist in the environment and water bodies. Some drugs like anti-epileptics, are persistent and some others are pseudo persistent. This means they are constantly being replaced while they degrade at reasonable rates. Some pharmaceuticals are highly fat soluble, which implies they will bio-accumulate. This way, they enter the cells and move up the food chains, thus their concentration increases within the process. When effluent is disinfected in the wastewater treatment plant, superbugs that breed in these plants release extracellular antibiotic resistance genes (ARG). The natural transformation method allows for the stable uptake, integration, and functional expression of extracellular DNA. This mechanism of horizontal gene transfer is considered to be responsible for antibiotic resistance and virulence factors. It's therefore necessary to make cell-free DNA non-functional by making wastewater inactivated using tested procedures.

This chapter summarizes the work done to enunciate threatening effects of pharmaceutical effluent and antimicrobial residues present within the pharmaceutical effluent on the ecology and environment. Efforts have also been made to enlist the available technologies for treating pharmaceutical effluent.

2. Influence of Antibiotic Residues on the Ecology

2.1. Evolutionary Consequences on Microbial Population

The evolutionary dynamics of a microbial population is influenced by antibiotic pollution, which results in a wide-range phenotypic response to antibiotics. Genetic or mutated adjustments in sensitive bacteria may occur due to presence of antibiotic residues within the microbial ecosystem. This allows the bacteria to live and grow further as antibiotic resistance bacteria (ARB) that carry antibiotic resistance genes (ARGs) [5]. Exposure to antibiotics may affect the developmental characteristics of microbes in many ways, such as genome reconstruction, mainly in the mobile genetic elements (MGEs) of the microbes, physiology of the microbiome, and triggering expression of a shock-response system in the microbes [6]. Bacterial populations exposed to low concentrations of antibiotics can cause bacterial lineages with increased phenotypic and genotypic diversity through the activation of an inherent survival response. It ends up in an increased mutation rate throughout the genome [7]. It has been reported that changes in the expression of varied genes lead to high phenotypic variability and, in some cases, increased virulence occurs due to the majority of antibiotics either binding on to the transcription factor or other regulatory mechanisms viz. riboswitches and quorum sensing [8]. Antibiotics may affect the enzymatic activity of bacteria (dehydrogenases, phosphatases, and ureases) which are considered important indicators of soil activity [9]. Thus, it leads to loss of important microbial taxonomy, which is very critical for ecology.

2.2. Adverse Effects of Antibiotics on Non-Microbial Population

Antibiotic pollution has a negative impact on vertebrates such as amphibians, reptiles, birds, and mammals in addition to microbial population. Liu et al. [10] observed improper locomotion behaviour, growth inhibition, cytotoxicity, and teratogenesis due to antibiotic pollution. Findings from a study conducted on *Xenopus tropicalis* revealed that antibiotic pollution within the environment is deadly to amphibians. The study also mentioned decreased body length, pericardial edema, and other malformations due to tetracycline exposure [10]. The presence of veterinary drug diclofenac in their food causes visceral gout and kidney disease and eventually death in Asian vulture populations [11]. Results of a study conducted to examine the acute toxic effects on a marine bacterium (*Vibrio fischeri*), a freshwater invertebrate (*Daphnia magna*), and Japanese Medaka fish (*Oryziaslatipes*), revealed that the LC_{50} for these organisms at two orders of enormity was higher than environmental concentrations [12]. Pharmaceutical residues also affect the bacteria, invertebrates, and algal populations adversely [13]. The release of antibiotics from the wastewater into the environment leads to the development of assorted resistant bacterial strains [14].

2.3. Adverse Effects on the Human Health

Toxicologists are facing new challenges due to the presence of complex mixtures of pharmaceutical residues in water. Antibiotics present in surface water and wastewater (between 0.01 and 1.0 μg/L) are the prospective sources for the transfer of novel antibiotic resistance mechanisms from environmental bacteria to humans and animal [15, 16]. Antibiotics available in potable water may not be entirely removed by traditional potable water treatment technologies and may persist for a long time [17, 18]. In wastewater treatment plants (WWTPs), bacteria from several sources and environments interact and exchange genes and are considered to be the major sources of ARB and ARGs which are released into the environment [19, 20].Thus, wastewater treatment plants have a vital role in the spread of resistance to antibiotics in humans [21, 22]. These antibiotics-resistant bacteria (ARB) and antibiotic-resistant genes (ARGs) may be transferred to humans [23]. Pathogenic ARBs could possibly reach into the tissues and cause acute reactions and communicable diseases. Testing for the consequences of a drug mixture at environmental levels on human cells revealed that a blend of pharmaceutical compounds slows down the expansion of embryonic kidney cells [24]. Pharmaceuticals triggered the stress-response signalling protein kinases (ERK1/2) and induced over expression of glutathione-S-transferase P1 gene. They observed morphological changes in HEK293 cells but found no evidence for apoptosis or necrosis. They also found that cells proliferation can be inhibited by a combination of medicines at ng/L levels by affecting their physiology and morphology.

2.4. Adverse Effects of Antibiotics on Plants

Kumari and Tripathi [25] evaluated pharmaceutical industry effluent toxicity by using genotoxocity associated with the use of *Allium cepa L.* (onion) root tip cells. Root tip cells were treated with effluent of different concentration and then genotoxicity was measured in terms of mitotic index (MI) and chromosomal aberrations in root tip cells. It was observed that root growth inhibition occurred at effluent concentration $\geq 50\%$ (v/v). The lowest MI (MI = 9.6%) was recorded in 25% of effluent concentration, which is considerably lower than the mitotic index with control (MI = 64.1%). Various c-mitosis, chromosome loss, chromosome break, micronucleated cells, and other chromosomal aberrations were observed in root tip cells after treatment with different concentrations of effluent. The result thus shows that the effluents contained highly toxic elements, which caused cytotoxic and genotoxic hazards.

2.5. Adverse Effects on Fish

Waste water released from WWTPs contains pharmaceutical effluents that when released to natural water bodies has an undesirable effect on the aquatic ecosystem, which results in disturbance of aquatic flora and fauna. Long-term effects like estrogen-induce vitellogenesis in male and high estrogenicity induced morbidity in fish were observed due to the constant release and exposure of pharmaceuticals to aquatic biota [26]. Continued low-level exposure to pharmaceuticals alters the species traits and behavior of aquatic organisms. The most common example of the shift in species trait is feminization of male fish because of the estrogen present in the aquatic environment [27]. Reduction in fish fecundity and adverse effect on reproductive capability of male and female fishes were observed when they were exposed to dutasteride [28].

Bordin et al. [29] studied drug-induced changes to quantify both the uptake and behavioural impact of a psychiatric drug on a predatory fish (*Percafluviatilis*) and its invertebrate prey (*Coenagrionhastulatum*). They observed that effects of pharmaceuticals on behaviour can differ between species. They found that behaviour of damselfly was unaffected but perch became more active. They also observed that on average 46% of the pharmaceutical accumulated in the predator came from ingested prey.

Corcoran et al. [30] reported adverse effects of the antibiotics available within the aquatic environment, especially on fish. Presence of tetracycline in the range of 0.1–50 μg/L within the environment acted as suppressant of the immune system of fish [31]. A study has revealed that the sub-inhibitory concentrations of macrolides influence the movement frequency of the embryo in zebra fish (*Danio rerio*) and induces deformity in the yolk sac edema, and uninflated swim bladder [32]. Growth and reproduction rate of zebra fish (*Danio rerio*) affected severely when they were exposed to norfloxacin and sulfamethoxazole at a concentration of 200 μg/L [33]. Antibiotics like fluoroquinolones and tetracyclines may also affect the behaviour and development of zebra fish (*Danio rerio*) by changing the expression of multiple genes [34].

The dose of most commonly used antidepressant, serotonin impact both physiology [35] and behaviour in many types of organisms, including fish [36, 37]. It is also important in activity, aggression and reproductive behaviour [38], which is negatively related to dose [39, 40]. It has been observed that antideprcssants reduce territorial aggression in coral reef fish [41] and locomotion and aggression in Siamese fighting fish [42)]. Fluoxetine reduced the feeding rate in both white and striped bass [43], as well as in goldfish [44]. Studies have also revealed that feeding rates of fathead minnow and hybrid striped bass reduced after exposure to 3.7 and 250 μg/L of fluoxetine and venlafaxine, respectively [45, 46]. Results of an experiment revealed that activity of Arabian killifish was drastically lowered when exposed to low concentrations of fluoxetine [47], and a similar effect was found for Siamese fighting fish [48]. The male fathead minnows showed unaffected reproductive behaviour by the antidepressants bupropion, fluoxetine, sertraline and venlafaxine, individually or in the form of mixture [49]. Whereas, the sociality of Arabian killifish increased after exposure to fluoxetine, and fathead minnows showed increased boldness after exposure to sertraline [50].

Increased activity in zebra fish [51] and pumpkinseed sunfish [52] has been recorded when they were exposed to one of the most commonly used psychiatric pharmaceuticals, diazepam at μg/L concentrations, and exposure to mg/L diazepam increased boldness in larval zebra fish [53]. Zebra fish when exposed to three benzodiazepines showed enhanced activity and affinity towards light [54]. Dominance in male fathead minnows increased when they were exposed to haloperidol, a drug that is used for the treatment of acute psychosis, aggression and acute delirium [55]. A study on perch from a natural population found that exposure to low doses (μg/L) of the diazepam, oxazepam increased activity and decreased sociality, and exposure to high concentrations of both drugs increased boldness [56].

Exposure of fathead minnows to μg/L of diphenhydramine (a kind of antihistamine) reduced their feeding rate, which is attributed to diphenhydramine's effect on serotonin levels [57]. In the same way, feeding rate and activity of Japanese medaka fish were reduced when exposed to carbamazepine, an antiepileptic drug, and diclofenac, a non-steroidal anti-inflammatory drug, separately [58]. Exposure of pumpkinseed sunfish to high mg/L neostigmine or pyrostigmine, cholinesterase inhibitors used to treat neuromuscular junction disorders, did not affect boldness. Pharmaceuticals with endocrine disrupting properties are another important group which affects aquatic life [59].

3. Pharmaceuticals Induced Antimicrobial Resistance

Antimicrobial Resistance (AMR) due to use of pharmaceuticals is a big challenge to modern medicine. All over the world, ordinary infections are becoming resistant to antimicrobial drugs which lead to multifaceted health issue with a huge impact on economic development. Rapidly growing pharmaceutical industries are mainly responsible for AMR as they work as a storehouse for antibiotics and antibiotic resistance genes (ARGs) in the environment (Figure 2).

Antimicrobial resistance, as a result of release of different classes of drugs into the environment, is a major health hazard. AMR take place when a microorganism evolves resistance against an antimicrobial agent. AMR is the ability of a microbe to resist the effects of drugs that could once successfully kill or inhibit the microbe. The main mechanisms involved in the development of antibiotic resistance are mutation [61] and acquisition of resistance genes [62] by horizontal gene transfer (HGT). Human, animal, and environment sectors play an important role in interrelated ecosystems in the development and transmission of AMR.

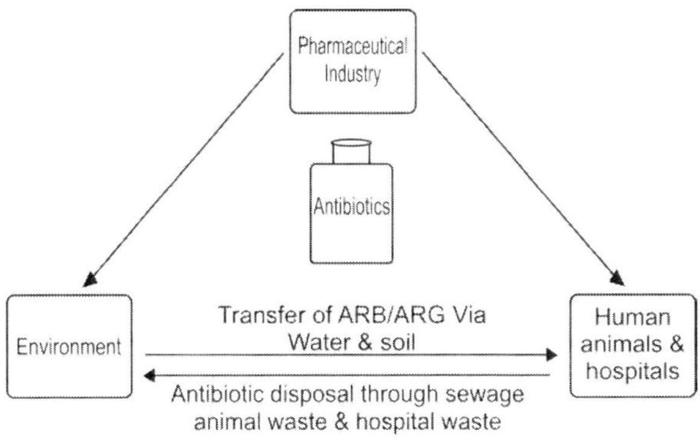

Figure 2. The interrelationship among pharmaceutical industry, environment, humans and animals. Adopted from [60].

Approximately 700,000 deaths each year globally occur on account of drug resistant infections. It has been estimated that in 2050, approximately 444 million people would be affected due to the inability to treat infections [63]. A continued rise in AMR will globally result in 2-3.5% reduction in GDP, with a cumulative cost of up to USD 100 trillion. The misuse and overuse of antibiotics is an important contributing factor of AMR. About 50% of the antibiotics prescribed for human use may be avoided. In the agriculture and aquaculture sectors, misuse of pharmaceuticals is still higher; they are mainly used as a growth promoter and as a substitute for good hygiene. The environment becomes a place of storage for resistant genes, as well as a place for the development and spread of resistance to pathogens. Improper use of antibiotics in humans, poultry, livestock, aquaculture and utilization of waste from antibiotic-fed animals as manure or fertilizer in fields, also add antibiotics and antibiotic-resistant genes (ARGs) in the water bodies and soil [64]. The perceptible sources which contribute to AMR's ecological role are agricultural and poultry farms, effluents from pharmaceutical industries and hospitals, and improperly disposed antibiotics [65]. The results of numerous scientific studies reveal the presence of antibiotic residues, antibiotic-resistant bacteria, and resistance genes in the environment [66, 67].

4. Presence of Pharmaceuticals in Water

Drugs consumed are first metabolized in the body into substances that are later on either excreted as unchanged pharmaceutical residues or its metabolites [68]. Wastewater treatment plants are not explicitly designed to remove pharmaceuticals from the effluent. For this reason, pharmaceuticals enter into lakes and rivers, and because of this; aquatic wildlife is persistently exposed to pharmaceuticals. Groundwater aquifers, which are main source of drinking water, may get polluted due to leaching of pharmaceuticals dissolved in wastewater. This means that most of the urban areas are likely to receive higher doses of pharmaceuticals. Rabiet et al., [69] reported presence of drugs like paracetamol, the anti-inflammatory diclofenac, and the epilepsy drug carbamazepine in many water bodies that are used for supply of potable water along the Lergue River in Southern France. They assessed the consequences of pharmaceuticals and other organic compounds present in treated wastewater. They also analysed water of wells used for supply of drinking water and effluents from two waste water treatment plants for the presence of pharmaceuticals. Common pharmaceutical compounds (amitryptilin, acetylsalicylic acid, carbamazepine, clenbuterol, diazepam, diclofenac, doxepin, gemfibrozil, ibuprofen, imipramine, ketoprofen, naproxen, nordiazepam, paracetamol, salbutamol, and terbutalin) and wastewater associated pollutants (caffeine, gadolinium anomaly, and boron) were detected in many water bodies used for supply of drinking water. The most frequently detected contaminants are paracetamol, caffeine, and diclofenac. The presence of paracetamol was high in concentrations (up to 11 μg/L and 211 ng/L, in a wastewater treatment plant effluent and in drinking water samples, respectively). In treated drinking water in Milan (Italy) clofibric acid, an active metabolite of many drugs used to lower cholesterol and diazepam, which is often prescribed for anxiety were detected [70]. Drinking water of Berlin

(Germany) contained clofibric acid, propylphenazone, and diclofenac [71]. Antibiotics in low concentrations were detected in municipal effluents and surface water [72, 73]. Kolpin et al. [74] found contaminants in the water of 80% sampled streams. The most frequently detected contaminants were the steroids coprostanol and cholesterol, the insect repellent *N,N*-diethyltoluamide (DEET), caffeine, and triclosan, an antimicrobial disinfectant. In most of the samples, the average number of compounds was seven, but the maximum number of compounds in a single sample was 38. The presence of anticancer drug tamoxifen and ciprofloxacin at a concentration of 25–38 and 7–103 ng/L respectively in river water has been reported [75]. In the effluent of membrane bioreactor WWTP presence of clarithromycin, metformin, atenolol, carbamazepine, and trimethoprim at the concentrations of >500 ng/L has been detected [76]. The concentrations of antidepressants, antibiotics, analgesic compounds, chemotherapy products, lipid regulators, and hormones in the environment range from 0.04 to 6.3 μg/L [77]. Release of hospital effluent into the municipal WWTP declined the biodegradation process of organic contaminant in WWTP because death of biodegrading microorganisms [78]. The addition of diclofenac continuously in anoxic sludge treatment process lowers the denitrifying capability of microbial community present in WWTP and reduces gas production [79]. Wilkinson et al. [80] presented results of a study of active pharmaceutical ingredient (API) pollution in 258 rivers across the globe, representing 137 geographic regions and environmental influence of 471.4 million people. They collected samples from 1,052 sites spread over 104 countries and analysed for 61 APIs. They observed highest cumulative API concentrations in South Asia, sub-Saharan Africa, and South America. Carbamazepine, metformin, and caffeine were the most frequently detected APIs, which were found at over half of the locations monitored. At least one API was found to be higher than concentrations considered safe for aquatic life at 25.7% of the sampling sites.

5. Ecological and Environmental Risks

The pharmaceutical industry has come out to be a new threat to environment due to its exponential growth [81]. These pharmaceutical pollutants pose a threat to the ecological balance in surface water, and public health. Antibiotic residues which enter in sewage are directly proportional to the quantity of antibiotics used in a particular area. After consumption, a huge percentage (approximately 90%) of the antibiotics is excreted unchanged through humans' and animals' urine and/or faeces [82]. These substances are not eradicated but only partially removed during sewage treatment [83, 84]. Dedicated on-site treatment at the source of pharmaceutical pollution is necessarily required before effluents are discharged to sewers and other water bodies. The first case of surface water contamination by antibiotics was reported in England in 1980, when one of the compounds from the macrolide, sulfonamide, and tetracycline group of antibiotics was detected in river water at a concentration of 1μg/L [85]. Antibiotic residues have also been detected in soil, hospital wastewaters, surface waters, groundwater, and drinking water [88-91]. The presence of a few commonly used antibiotics in a sewage treatment plant (STP) and the environmental concentration of these antibiotics in the Yamuna River at Delhi were detected [92]. They observed concentration of ampicillin in wastewater influents in the range of 104.2 ± 98.11 μg/L and effluents in the range of 12.68 ± 8.38 μg/L, respectively. Fluoroquinolones and cephalosporins concentrations in wastewater were very low. Significant quantities of antibiotics were released and detected in effluents as the treatment efficiencies varied between 55 and 99%. Larsson et al. [93] observed very high concentrations of ciprofloxacin residues ranging up to 31,000 μg/L in the effluent of wastewater treatment plant (WWTP) at Patancheru, Hyderabad. Kaveri River, Tamiraparani River, and Vellar River drawing treated sewage and industrial effluents, had higher concentration of pharmaceutical residues [94].

Antibiotic residues present even at trace levels are distressing the ecosystem's sustainability. There is no information on the direct effects of contaminated water on humans, but the effects on other organisms are well documented. Death and a drop in reproduction of standard test organisms like *V. fischeri, D. magna, M. macrocopa, O. latipes,* and some invertebrates have been reported due to pharmaceutical pollution [95, 96]. Alighardashi et al. [97] reported intense sensitivity of sludge bacteria to erythromycin causing flock breakdown and breakage of filaments due to pharmaceutical pollution.

Algal species are very sensitive to antibiotics, and, as algae are the basis of the food chain, even minor decreases in algae population affect the nutrient cycle and food supply. Effect concentration (EC) of organisms decreases when they are exposed to a mixture of drugs. DeLorenzo and Fleming [98] reported that EC_{50} reduced by 75% in Dunaliellatertiolecta for a mixture of pharmaceuticals when compared to the same level of single pharmaceuticals. The presence of pharmaceutical residues in wastewater has been linked to the development of antibiotic resistant bacteria [99, 100] and the threat to aquatic organisms [101].

Substances such as synthetic female steroid hormones and natural estrogenic hormones are biologically very strong compounds. With the conventional wastewater treatment, these hormones can only be eliminated partially. This is why they have been detected in the effluent of various European wastewater treatment plants [102, 103]. For example, four of five metabolites and 27 of 32 pharmaceutical substances were detected in municipal wastewater treatment plant effluents. Peak values of over 1μg/L were measured in surface waters e.g., beta-blockers, antiepileptic drugs, lipid regulating agents, and antiphlogistics [104]. Over the last 15 years period, a 50% decline in fish catch has been observed due to organic micropollutants in several rivers and streams in Switzerland [105, 106].

Bielena et al. [107] performed ecotoxicological, chemical, and microbiological analyses of effluents from two pharmaceutical industries in Croatia. They observed that the total concentration of azithromycin (AZI) and two macrolide by-products from its synthesis was up to 638 μg/L in treated effluent of the company synthesizing macrolide antibiotic AZI. As a result, the total concentration in the recipient river water was up to 30 μg/L. Effluents from another company producing veterinary antibiotics contained fluoroquinolones, trimethoprim, sulfonamides and tetracyclines in concentrations of up to 200μg/L.In the studied effluents, a high frequency of culturable bacteria resistant to AZI (up to 83%) or sulfamethazine (up to 90%) and oxytetracycline (up to 50%) were also found. It was observed that toxicity to algae and water fleas generally exceeded the permissible limits. Highly contaminated effluents induced various deformities in zebra fish embryos.

6. Removal of Pharmaceutical Pollutants

Active pharmaceutical compounds and other wastewater elements from effluents cannot be completely removed using traditional wastewater treatment methods, such as activated sludge. Pharmaceuticals pose a great threat in the area of environmental remediation. Therefore, to mitigate their impact and reduce their consequences on the environment and human health, alternative add-on systems for traditional wastewater treatment plants are being developed continuously. Diverse methods (chemical, physical, and biological) have been used to get rid of antibiotics from water and wastewater. Pharmaceutical pollutants can be removed from water and wastewater using adsorption techniques such as carbon nanotubes, graphene oxide, activated carbon [108], biological degradation processes, chemical advanced oxidation processes such as fenton, photocatalytic oxidation, photophenton, ionizing irradiation, ozonation, [109, 110, 111, 112], Membrane Bio-Reactor (MBR), algae photobioreactor, constructed wetland, and stabilization ponds [113, 114, 115, 116]. Use of biological treatment methods and advanced oxidation processes in combination has resulted in higher removal efficiency for pharmaceutical products [109]. Persistent pharmaceutical pollutants (PPPs) can be completely decomposed with advanced oxidation technologies (AOTs) such as dielectric barrier discharges (DBDs) based on free radical mechanisms. These AOTs have the potential to be useful in the decomposition of biologically intractable and persistent organic pollutants found in wastewater [117, 118]. Oxidation can easily be accomplished using hydroxyl radicals as primary non-selective oxidants [119]. Most commonly used AOTs are UV disinfection (photolysis) and a variety of photolytic combinations including UV/H_2O_2, UV/O_3, UV/H_2O_2/O_3, wet air oxidation, electron beam, photocatalysis, plasma technologies, supercritical water oxidation, ultrasound cavitation, water sonolysis, electrochemical advanced oxidation processes, microwave plasma-based AOTs [120, 121, 122, 123]. Because of their plasma properties and reactor configurations, AOTs stimulated by dielectric barrier discharge (DBD) are extremely effective [124, 125, 126]. The elimination of pharmaceuticals in batch experiments with sand under natural aerobic and anoxic conditions was not significant [127]. Flocculation using iron chloride was also found ineffective for elimination of the selected pharmaceuticals (clofibric acid, bezafibrate, diclofenac, and carbamazepine,). The concentrations of diclofenac and carbamazepine could be decreased by more than 90% with ozonation at a concentration of 0.5 mg/L ozone, while bezafibrate was reduced by 50% with a 1.5 mg/L ozone dose. Clofibric acid could not be removed even at the ozone concentration of 3 mg/L.

The membrane bioreactor (MBR) combines a low-pressure membrane with conventional activated sludge treatment (CAS). This system is more efficient at removing micro-pollutants as compared to conventional activated sludge process because less sludge is produced and very little suspended particles are present in the filtrate [128]. MBR technology removes a significant amount of organic pollutants due to the long sludge retention time and efficient nitrogen removal by slow growing autotrophic bacteria. The Membrane Bioreactor is a better treatment process than CAS because it retains hydrophobic compounds and also slow growing nitrifying microorganism within the reactor [129]. Work conducted by Wei et al., [130] revealed that for decreasing membrane fouling rate and for removal of highly persistent micro-pollutant; presence of supporting medium for microbial growth in MBR would be a valuable technique. Use of MBR is a prevalent practice in hospital wastewater treatment. A study reported 99% elimination of 17ß-ethinylestradiol in nitrifier-enriched biomass of MBR [131]. Dawas et al. [132] observed that pharmaceutical residues degradation is accelerated due to high nitrifying activity. When the compounds like caffeine, acetaminophen, sulfamethoxazole, carbamazepine, and gemfibrozil passed through MBR, their concentrations reduced with removal efficiencies varying between 99.1% for sulfamethoxazole and 99.9% for acetaminophen [133]. Gemfibrozil could not be removed by conventional activated sludge treatment, whereas the MBR could remove in the range of 30–40%. With the use of MBR technology 81% of sulfamethoxazole could be removed whereas by the conventional activated sludge method only 75% could be removed. Kimura et al. [134] observed removal efficiency of ibuprofen, clofibric acid, mefenamic acid and diclofenac were the same in both CAS and MBR but for ketoprofen and naproxen removal efficiency was higher in MBR system. MBR system is more efficient than CAS treatment for removing the persistent micro-pollutant particularly the not readily degradable compounds. Bernhard et al. [135] observed that MBR process can more efficiently remove polar compounds like sulfophenyl carboxylate, diclofenac, and mecoprop. Joss et al., [136] revealed that increase in retention time in MBR enhanced the degradation of estrogen. Radjenović et al. [137] observed that degradation efficiency of pharmaceutical compounds like metoprolol, diclofenac, and clofibric acid was 58.7%, 87.4%, and 71.8% respectively in MBR whereas in CAS treatment, it was only 50% and 27% for diclofenac and clofibric acid respectively. The removal efficiency of ibuprofen was the same for both treatment processes. Submerged attached bio-filter had higher removal efficiency of gemfibrozil and ciprofloxacin as compared to MBR [138]. Use of activated carbon in MBR absorbs low molecular weight contaminants and provides support for the attached bacterial growth. Degradation of pharmaceutical compounds such as diclofenac, diazepam, and carbamazepine were increased by adding 0.5 g/L granular activated carbon (GAC) into the aeration tank of activated sludge [139]. Addition of high concentration of powdered activated carbon (PAC) (1 g/L) in MBR enhances the carbamazepine removal up to 90% [140]. MBR after PAC addition resulted in high removal of diazepam and carbamazepine and profusion of Accumulibacterphosphatis and Nitrosomonas [141]. However, activated carbon in MBR must be used with utmost care as overdose may result in high membrane fouling, increase the viscosity of sludge and therefore reduced sludge drying.

7. Pharmaceuticals Removal Efficiency

The efficiency of pharmaceutical removal depends on its nature and the effectiveness of treatments, which varies widely depending on the type of drug. Hydrophobic compounds can be effectively removed by oxidizing them with free chlorine. A number of hydrophilic compounds can partially be removed through adsorption processes. Waste water treatment systems, together with ozonation and activated carbon treatments, remove the majority of the pharmaceuticals. Increased elimination efficiency for a range of pharmaceutical substances can be achieved by combining these treatments with other traditional procedures. Pharmaceutical removal efficiencies are greatly affected by technologies used in removal plants, type of operation, geographic location, and environmental factors. Biological treatment reduced dissolved fluoroquinolone (ciprofloxacin and norfloxacin) concentrations by 40–60% [142]. McArdell et al. [143] reported limited elimination of macrolides in effluents from wastewater treatment plants [143]. Varied removal efficiencies among WWTPs have been reported; they are high (72–100%) for estrogens, non-steroidal anti-inflammatory drugs, and caffeine, but some antibiotic groups viz. macrolides, penicillin, and imidazole were not removed [144]. Pharmaceutical removal efficiencies were lower (<50%) for antibiotics in sewage treatment plants [145]. Thus, effluent may contain numerous compounds and all of these compounds cannot be removed with any of the treatment process.

Conclusion

Pharmaceuticals that are used for human and veterinary treatment can get into the aquatic ecosystem through sewage and wastewater treatment plants. There is still a lack of knowledge in understanding the interaction between pharmaceutical metabolites and resistance promotion. The understanding of the synergistic effects caused by organisms being exposed to many substances simultaneously in receiving water bodies is also lacking. Much research work is required to assess the potential adverse effects of pharmaceutical pollution on human, ecology and environment. Advanced treatment technologies need to be evolved for removing the xenobiotics in such a way that they can reduce hazardous impacts on the environment. Antibiotics contain those chemicals which retard the activities of sewage bacteria, seriously affecting the rate of decomposition of organic matter. As per the above discussion, it is also evident that antibiotics are also toxic to nitrifying bacteria in the wastewater treatment process. In conclusion, effluents discharged from pharmaceutical industries can pose significant ecological and public health hazards due to their toxicity to the environment and aquatic life and create risks for development and spread of antibiotic resistance.

References

[1] Tiwari, B., Sellamuthu, B., Ouarda, Y., Drogui, P., Tyagi, R.D., & Buelna, G. (2017). "Review on fate and mechanism of removal of pharmaceutical pollutants from wastewater using biological approach." *Bioresource Technology*, 224: 1-12. https://doi.org/10.1016/j.biortech.2016.11.042.

[2] Cardoso, O., Porcher, J., & Sanchez, W. (2014). "Factory-discharged pharmaceuticals could be a relevant source of aquatic environment contamination: Review of evidence and need for knowledge." *Chemosphere,* 115 : 20-30. doi: 10.1016/j.chemosphere.2014.02.004.

[3] Burns, E.E., Carter, L. J., Nape, J., Thomas-Oates, J., & Boxall, A. B. A. (2018). "Application of prioritization approaches to optimize environmental monitoring and testing of pharmaceuticals." *Journal of Toxicology and Environmental Health, Part B*, 21, 3: 115-141, doi: 10.1080/10937404.2018.1465873.

[4] Küster, A., & Adler, N. (2014), "Pharmaceuticals in the environment: Scientific evidence of risks and its regulation." *Philosophical Transactions of the Royal Society B: Biological Sciences*, 369: 1656. http://dx.doi.org/10.1098/rstb.2013.0587.

[5] Martinez, J. L. (2009). "The role of natural environments in the evolution of resistance traits in pathogenic bacteria." *Proc Biol Sci*, 276: 2521–2530. https://doi.org/10.1098/rspb.2009.0320.

[6] Martínez, J. L. (2017). "Effect of antibiotics on bacterial populations: a multi-hierarchical selection process." *F1000Research, 6(F1000:Faculty Rev) 51.* https://doi.org/10.12688/f1000research.9685.1.

[7] Foster, P. L. (2007). "Stress-induced mutagenesis in bacteria." *Crit Rev Biochem Mol Biol,* 42: 373–397. https://doi.org/10.1080/10409230701648494.

[8] Davies, J., Spiegelman, G.B., Yim, G. (2006). "The world of sub inhibitory antibiotic concentrations." *Curr Opin Microbiol,* 9: 445–453. https://doi.org/10.1016/j.mib.2006.08.006.

[9] Cycon, M., Mrozik, A., Piotrowska-Seget, Z. (2019). "Antibiotics in the soil environment degradation and their impact on microbial activity and diversity." *Front Microbiol,* 10: 338. https://doi.org/10.3389/fmicb.2019.00338.

[10] Liu L., Wu W., Zhang J., Lv P., Xu L., Yan, Y. (2018) "Progress of research on the toxicology of antibiotic pollution in aquatic organisms." *Acta Ecol Sin,* 38, 1 : 36-41. https://doi.org/10.1016/j.chnaes.2018.01.006.

[11] Oaks, J. L., Gilbert, M., Virani, M.Z., Watson, R.T., Meteyer, C.U., Rideout, B.A., Shivaprasad, H., Ahmed, S., Chaudhry, M. J. I., & Arshad, M. (2004). "Diclofenac residues as the cause of vulture population decline in Pakistan." *Nature*, 427: 630-633.

[12] Kim, Y., Jung, J., Oh, S. R., & Choi, K. (2008). "Aquatic toxicity of cartap and cypermethrin to different life stages of Daphnia magna and Oryzias latipes." *Journal of Environmental Science and Health. Part. B*, 43: 56–64. doi: 10.1080/03601230701735029.

[13] Hernando, M. D., Mezcua, M., Fernandez-Alba, A. R., & Barcelo, D. (2006). "Environmental risk assessment of pharmaceutical residues in wastewater effluents, surface waters, and sediments." *Talanta*, 69, 334–342. doi: 10.1016/j.talanta.2005.09.037.

[14] Kummerer, K. (2004). "Resistance in the environment." *Journal of Antimicrobial Chemotherapy*, 54, 2: 311–320. https://doi.org/10.1093/jac/dkh325.

[15] Aslan, A., Cole, Z., Bhattacharya, A., & Oyibo, O. (2018) "Presence of antibiotic-resistant Escherichia coli in wastewater treatment plant effluents utilized as water reuse for irrigation." *Water,* 10, 6 : 805 . https://doi.org/10.3390/w10060805.

[16] Manaia, C. M., Macedo, G., Fatta-Kassinos, D., Nunes, O. C. (2016). "Antibiotic resistance in urban aquatic environments: can it be controlled?" *Appl Microbiol Biotechnol,* 100 :1543-1557. https://doi.org/10.1007/s00253-015-7202-0.

[17] Ben, Y., Fu, C., Hu, M., Liu, L., Wong, M. H., & Zheng, C. (2019). "Human health risk assessment of antibiotic resistance associated with antibiotic residues in the environment: a review." *Environ Res,* 169: 483–493. https://doi.org/10.1016/j.envres.2018.11.040.

[18] Wang, Q. J., Mo, C. H., Li, Y. W., Gao, P., Tai, Y. P., Zhang, Y., Ruan,Z. L., Xu, J. W. (2010). "Determination of four fluoroquinolone antibiotics in tap water in Guangzhou and Macao." *Environ Pollut*, 158: 2350–2358. doi: 10.1016/j.envpol.2010.03.019.

[19] Alexander, J., Hembach, N, Schwartz T (2020) Evaluation of antibiotic resistance dissemination by wastewater treatment plant effluents with different catchment areas in Germany. *Sci Rep* 10:1–9. https://doi.org/10.1038/s41598-020-65635-4.

[20] Karkman, A., Do, T. T., Walsh, F., & Virta, M. P. J. (2017). "Antibiotic-resistance genes in wastewater." *Trends Microbiol*, 26: 220–228. https://doi.org/10.1016/j.tim.2017.09.005.

[21] Bouki, C., Venieri, D., & Diamadopoulos, E. (2013). "Detection and fate of antibiotic resistant bacteria in wastewater treatment plants: a review." *Ecotoxicol Environ Saf,* 91:1–9. https://doi.org/10.1016/j.ecoenv.2013.01.016.

[22] Rizzo, L., Manaia, C., Merlin, C., Schwartz, T., Dagot, C., Ploy, M. C., Michael, I., Fatta-Kassinos, D. (2013). "Urban wastewater treatment plants as hotspots for antibiotic resistant bacteria and genes spread into the environment: a review." *Sci Total Environ,* 447: 345–360. doi: 10.1016/j.scitotenv.2013.01.032.

[23] Manaia, C. M. (2017). "Assessing the risk of antibiotic resistance transmission from the environment to humans: non-direct proportionality between abundance and risk." *Trends Microbiol*, 25, 3: 173–181. doi: 10.1016/j.tim.2016.11.014.

[24] Pomati, F., Castiglioni, S., Zuccato, E., Fanelli, R., Vigetti, D., Rossetti, C., & Calamari, D. (2006). "Effects of a complex mixture of therapeutic drugs at environmental levels on human embryonic cells." *Environ Sci Technol*, 40, 7: 2442-2447. doi: 10.1021/es051715a.

[25] Kumari, V., & Tripathi, A.K. (2019). "Characterization of pharmaceuticals industrial effluent using GC–MS and FT-IR analyses and defining its toxicity." *Appl Water Sci*, 9: 185. https://doi.org/10.1007/s13201-019-1064-z.

[26] Jukosky, J. A., Watzin, M.C., & Leiter, J. C. (2008). "The effects of environmentally relevant mixtures of estrogens on Japanese medaka (Oryziaslatipes) reproduction." *Aquat. Toxicol*, 86, 2: 323-331. doi: 10.1016/j.aquatox.2007.11.012.

[27] Gross-Sorokin, M.Y., Roast, S. D., & Brighty, G.C. (2005). "Assessment of feminization of male fish in English rivers by the Environment Agency of England and Wales." *Environ. Health Perspect*, 114, S-1: 147-151. doi: 10.1289/ehp.8068.

[28] Casaluci, M. L., Hannah, R. E., Sumpter, J. P. (2013). "Mode of action of human pharmaceuticals in fish: the effects of the 5-alpha-reductase inhibitor, dutasteride, on reproduction as a case study. Aquat." *Toxicol*, 128–129. https://www.academia.edu/60937401/Mode_of_action_of_human_pharmaceuticals_in_fish_The_effects_of_the_5_alpha_reductase_inhibitor_dutasteride_on_reproduction_as_a_case_study.

[29] Brodin, T., Piovano, S., Fick, J., Klaminder, J., Ynen, M. & Jonsson, M. (2014). "Ecological effects of pharmaceuticals in aquatic systems—impacts through behavioural alterations." *Phil. Trans. R. Soc. B*, 369: 2013058. http://doi.org/10.1098/rstb.2013.0580.

[30] Corcoran, J., Winter, M. J., & Tyler, C. R. (2010). "Pharmaceuticals in the aquatic environment: a critical review of the evidence for health effects in fish." *Crit Rev Toxicol*, 40: 287–304. https://doi.org/10.3109/10408440903373590.

[31] Grondel, J. L., Gloudemans, A. G., Van-Muiswinkle, W. B. (1985). "The influence of antibiotics on the immune system II. Modulation of fish leukocyte responses in culture." *Vet Immunol Immunopathol*, 9: 251–260. https://doi.org/10.1016/0165-2427(85)90075-3.

[32] Wang, H., Che, B., Duan, A., Mao, J., Dahlgren, R. A., Zhang, M., Zhang, H., Zeng, A., & Wang, X. (2014) "Toxicity evaluation of β-diketone antibiotics on the development of embryo-larval zebrafish (Danio rerio)." *Environ Toxicol*, 29: 1134–1146. https://doi.org/10.1002/tox.21843.

[33] Yan, Z., Lu, G., Ye, Q., & Liu, J. (2016). "Long-term effects of antibiotics, norfloxacin, and sulfamethoxazole, in a partial life-cycle study with zebrafish (Danio rerio): effects on growth, development, and reproduction. *Environ Sci Pollut Res* 23:18222–18228. https://doi.org/10.1007/s11356-016-7018-1.

[34] Zhang, Y., Wang, X., Yin, X., Shi, M., Dahlgren, R. A, & Wang, H. (2016). "Toxicity assessment of combined fluoroquinolone and tetracycline exposure in zebrafish (Danio rerio)." *Environ Toxicol,* 31: 736–750. https://doi.org/10.1002/tox.22087.

[35] Meston, C.M., & Frohlich, P.F. (2000) "The neurobiology of sexual function." *Arch. Gen. Psychiatry*, 57: 1012–1030. doi: 10.1001/archpsyc.57.11.1012.

[36] Gunnarsson, L., Jauhiainen, A., Kristiansson, E., Nerman, O., & Larsson, D.G.J. (2008). "Evolutionary conservation of human drug targets in organisms used for environmental risk assessments." *Environ. Sci. Technol.*, 42: 5807–5813. doi: 10.1021/es8005173 .

[37] Kreke, N., & Dietrich, D.R. (2008). "Physiological endpoints for potential SSRI interactions in fish." *Crit. Rev. Toxicol,* 38: 215–247. doi:10.1080/1040844080189 1057.

[38] Lillesaar, C. (2011). "The serotonergic system in fish." *J. Chem. Neuroanat.*, 41: 294–308. doi: 10.1016/j.jchemneu.2011.05.009.

[39] Lesch, K. P. & Merschdorf, U. (2000). "Impulsivity, aggression, and serotonin: a molecular psychobiological perspective." *Behav. Sci. Law,* 18: 581–604. doi:10.1002/1099-0798(200010)18:5<581::aid-bsl411>3.0.co;2-l.

[40] Nelson, R.J., & Chiavegatto, S. (2001). "Molecular basis of aggression." *Trends Neurosci.*, 24, 713–719. doi: 10.1016/S0166-2236(00)01996-2.

[41] Perreault, H.A.N., Semsar, K., & Godwin, J. (2003). "Fluoxetine treatment decreases territorial aggression in a coral reef fish." *Physiol. Behav.*, 79: 719–724. doi: 10.1016/s0031-9384(03)00211-7.

[42] Kohlert, J. G., Mangan, B. P., Kodra, C., Drako, L., Long, E. & Simpson, H. (2012). "Decreased aggressive and locomotor behaviors in Betta splendens after exposure to fluoxetine." *Psychol. Rep.*, 110: 51–62. doi: 10.2466/02.13.PR0.110.1.51-62.

[43] Gaworecki, K. M., & Klaine, S. J. (2008). "Behavioral and biochemical responses of hybrid striped bass during and after fluoxetine exposure." *Aquat. Toxicol.* 88: 207–213. doi: 10.1016/j.aquatox.2008.04.011.

[44] Mennigen, J. A., Sassine, J., Trudeau, V. L. & Moon, T.W. (2010). "Waterborne fluoxetine disrupts feeding and energy metabolism in the goldfish *Carassius auratus*." *Aquat. Toxicol.*, 100: 128–137. doi: 10.1016/j.aquatox.2010.07.022.

[45] Stanley, J. K., Ramirez, A. J., Chambliss, C. K., & Brooks, B.W. (2007). "Enantiospecific sublethal effects of the antidepressant fluoxetine to a model aquatic vertebrate and invertebrate." *Chemosphere*, 69: 9–16. doi: 10.1016/j.chemosphere.2007.04.080.

[46] Bisesi, J. H. J., Bridges, W., & Klaine, S.J. (2014). "Effects of the antidepressant venlafaxine on fish brain serotonin and predation behavior." *Aquat. Toxicol.*, 148: 130–138. doi:10.1016/j.aquatox.2013.12.033.

[47] Barry, M.J. (2013). "Effects of fluoxetine on the swimming and behavioural responses of the Arabian killifish." *Ecotoxicology*, 22: 425–432. doi: 10.1007/s10646-012-1036-7.

[48] Lynn, S. E., Egar, J. M., Walker, B. G., Sperry, T. S., &Ramenofsky., M. (2007). "Fish on Prozac: a simple, noninvasive physiology laboratory investigating the mechanisms of aggressive behavior in Betta splendens." *Adv. Physiol. Educ.*, 31: 358–363. doi: 10.1152/advan.00024.2007.

[49] Schultz, M. M., Painter, M. M., Bartell, S. E., Logue, A., Furlong, E. T., Werner, S. L. & Schoenfuss, H. L. (2011). "Selective uptake and biological consequences of environmentally relevant antidepressant pharmaceutical exposures on male fathead minnows." *Aquat. Toxicol.*, 104: 38–47. doi: 10.1016/j.aquatox.2011.03.011.

[50] Valenti, T. W., Gould, G. G., Berninger, J. P., Connors, K. A., Keele, N. B., Prosser, K. N., & Brooks, B.W. (2012). "Human therapeutic plasma levels of the selective serotonin reuptake inhibitor (SSRI) sertraline decrease serotonin reuptake transporter binding and shelter-seeking behaviour in adult male fathead minnows" *Environ. Sci. Technol.*, 46: 2427–2435. doi: 10.1021/es204164b.

[51] Oggier, D. M., Weisbrod, C. I., Stoller, A.M., Zenker, A.K., & Fent, K. (2010). "Effects of diazepam on gene expression and link to physiological effects in different life stages in zebrafish Danio rerio." *Environ. Sci. Technol,* 44: 7685–7691. doi: 10.1021/es100980r.

[52] Brandao, F. P., Rodrigues, S., Castro, B.B., Goncalves, F., Antunes, S. C., & Nunes, B. (2013). "Short-term effects of neuroactive pharmaceutical drugs on a fish species: biochemical and behavioural effects." *Aquat. Toxicol.*, 144–145. doi: 10.1016/j.aquatox.2013.10.005.

[53] Richendrfer, H., Pelkowski, S. D., Colwill, R. M. and Creton R. 2012. On the edge: pharmacological evidence for anxiety-related behavior in zebrafish larvae. *Behav. Brain Res.*, 228: 99–106. doi: 10.1016/j.bbr.2011.11.041.

[54] Gebauer, D. L., Pagnussat, N., Piato, A. L., Schaefer, I. C., Bonan, C. D. & Lara, D. R. (2011). "Effects of anxiolytics in zebrafish: similarities and differences between benzodiazepines, buspirone and ethanol." *Pharmacol. Biochem. Behav.*, 99: 480–486. doi: 10.1016/j.pbb.2011.04.021.

[55] Villeneuve, D. L. (2010). "Effects of a dopamine receptor antagonist on fathead minnow dominance behavior and ovarian gene expression in the fathead minnow and zebrafish. Ecotoxicol." *Environ. Saf.*, 73: 478–485. doi: 10.1016/j.ecoenv.2009.09.018.

[56] Brodin, T., /Fick, J., Jonsson, M., & Klaminder, J. (2013). "Dilute concentrations of a psychiatric drug alter behavior of fish from natural populations." *Science,* 339: 814–815. doi: 10.1126/science.1226850.

[57] Berninger, J. P., Du, B., Connors, K. A., Eytcheson, S. A, Kolkmeier, M. A., Prosser, K. N., Valenti, T. W., Chambliss, C. K. & Brooks, B.W. (2011). "Effects of the antihistamine diphenhydramine on selected aquatic organisms. Environ." *Toxicol. Chem.*, 30: 2065–2072. doi: 10.1002/etc.590.

[58] Nassef, M., Matsumoto, S., Seki, M., Khalil, F., Kang, I. J., Shimasaki, Y., Oshima, Y., & Honjo, T. (2010). "Acute effects of triclosan, diclofenac and carbamazepine on feeding performance of Japanese medaka fish (*Oryzias latipes*)." *Chemosphere*, 80: 1095–1100. doi: 10.1016/j.chemosphere.2010.04.073.

[59] Soeffker, M., & Tyler, C. R. (2012). "Endocrine disrupting chemicals and sexual behaviors in fish- a critical review on effects and possible consequences." *Crit. Rev. Toxicol.*, 42: 653–668. doi: 10.3109/10408444.2012.692114.

[60] Kotwani, A., Joshi, J., & Kaloni, D. (2021). "Pharmaceutical effluent: a critical link in the interconnected ecosystem promoting antimicrobial resistance." *Environ Sci Pollut Res,* 28: 32111–32124. https://doi.org/10.1007/s11356-021-14178-w.

[61] Martinez, J. L., & Baquero, F. (2000). "Mutation frequencies and antibiotic resistance. Antimicrob." *Agents Chemother*, 44: 1771–1777. doi: 10.1128/AAC.44.7.1771-1777.2000.

[62] Davies, J. (1994). "Inactivation of antibiotics and the dissemination of resistance genes." *Science.*, 264: 375–382. doi: 10.1126/science.8153624.

[63] O'Neill, J. (2016). *"The Review on Antimicrobial Resistance-Tackling Drug-Resistant Infections Globally: Final Report and Recommendations."* https://amr-review.org/sites/default/files/160518_Finalpaper_withcover.pdf.

[64] Meersche, T.V.D., Rasschaert, G., Nest, T.V., Haesebrouck, F., & Herman, L. (2020) "Longitudinal screening of antibiotic residues, antibiotic resistance genes and zoonotic bacteria in soils fertilized with pig manure." *Environ Sci Pollut Res*, 28016–28029. https://doi.org/10.1007/s11356-020-09119-y.

[65] Khurana, A., Sinha, R. (2019). "Tackling AMR Starts with the environment." *Sanit Environ,* 78–80. http://resistancecontrol.info/wp-content/uploads/2019/05/Khurana.pdf.

[66] Kumar, R., & Kotwani, A. (2017). "Rising threat of antimicrobial resistance: judicious use of antibiotics is the way forward." *Indian J Chest Dis Allied Sci*, 59: 65–166. https://www.ncbi.nlm.nih.gov/pmc/articles/PMC8086231/.

[67] Singer, A. C., Shaw, H., Rhodes, V., & Hart, A. (2016). "Review of antimicrobial resistance in the environment and its relevance to environmental regulators." *Front Microbiol*, 7. https://doi.org/10.3389/fmicb.2016.01728.

[68] Fatta-Kassinos, D., Meric, S., & Nikolaou, A. (2011). "Pharmaceutical residues in environmental waters and wastewater: current state of knowledge and future research." *Anal Bioanal Chem.,* 399: 251–275. https://doi.org/10.1007/s00216-010-4300-9.

[69] Rabiet, M., Togola, A., Brissaud, F., Seidel, J. L., Budzinski, H., & Elbaz-Poulichet, F. (2006). "Consequences of treated water recycling as regards pharmaceuticals and drugs in surface and ground waters of a medium-sized Mediterranean catchment." *Environ Sci Technol*. 40, 17: 5282-8. doi: 10.1021/es060528p.

[70] Zuccato, E., Calamari, D., Natangelo, M., & Fanelli, R. (2000). "Presence of therapeutic drugs in the environment." *Lancet.*, 355: 9217:1789-90. doi: 10.1016/S0140-6736(00)02270-4.

[71] Heberer, T. (2002). "Tracking persistent pharmaceutical residues from municipal sewage to drinking water." *Journal of Hydrology*, 266,(3-4):175-189. doi: 10.1016/S0022-1694(02)00165-8.

[72] Alder, A. C. (2002). "Environmental exposure and risk assessment of fluoroquinolone antibacterial agents in wastewater and river water of the Glatt Valley Watershed, Switzerland." *Environ Sci Technol,* 36:3645–3651. https://doi.org/10.1021/es0256212.

[73] Wiegel, S., Aulinger, A., & Brockmeyer, R. (2004). "Pharmaceuticals in the river Elbe and its tributaries." *Chemosphere,* 57: 107–126. https://doi.org/10.1016/j.chemosphere.2004.05.017.

[74] Kolpin, D.W., Furlong, E. T., Meyer, M. T., Thurman, E.M., Zaugg, S.D., Barber, L. B., and Buxton, H. T. (2002). "Pharmaceuticals, hormones, and other organic wastewater contaminants in U.S. streams, 1999-2000: a national reconnaissance." *Environ Sci Technol,* 15, 36, (6): 1202-11. doi: 10.1021/es011055j .

[75] Ferrando-Climent, L., Rodriguez-Mozaz, S., & Barceló, D. (2014). "Incidence of anticancer drugs in an aquatic urban system: from hospital effluents through urban wastewater to natural environment." *Environ. Pollut.,* 193: 216-223. doi: 10.1016/j.envpol.2014.07.002.

[76] Kim, M., Guerra, P., Shah, A., Parsa, M., Alaee, M., & Smyth, S. A. (2014). "Removal of pharmaceuticals and personal care products in a membrane bioreactor wastewater treatment plant." *Water Sci. Technol.*, 69, 11: 2221-9. doi: 10.2166/wst.2014.145.

[77] Jones, O.A.H., Voulvoulis, N., Lester, J. N. (2001). "Human pharmaceuticals in the aquatic environment a review." *Environ. Technol.*, 22, 12: 1383-1394. doi: 10.1080/09593332208618186.

[78] Pauwels, B., Verstraete, W. (2006). "The treatment of hospital wastewater: an appraisal." *J. Water Health*, 4, 4: 405-416. https://pubmed.ncbi.nlm.nih.gov/17176811/.

[79] Ozdemir, G., Aydin, E., Topuz, E., Yangin-Gomec, C., & Tas, D.O. (2015). "Acute and chronic responses of denitrifying culture to diclofenac." *Bioresour. Technol.*, 176: 112-4. doi: 10.1016/j.biortech.2014.10.055.

[80] Wilkinson, J. L., Boxall, A. B. A., Kolpin, D. W., & Leung, K.M.Y. (2022). "Pharmaceutical pollution of the world's rivers." In *Proceedings of the National Academy of Sciences.* 119, 8: https://doi.org/10.1073/pnas.2113947119.

[81] Khan, A.H., Aziz, H.A., Khan, N.A. (2022). "Impact, disease outbreak and the eco-hazards associated with pharmaceutical residues: a Critical review." *Int. J. Environ. Sci. Technol.* 19, 677–688. https://doi.org/10.1007/s13762-021-03158-9.

[82] Hirsch, R., Ternes, T. A., Haberer, K., Mehlich, A., Ballwanz, F., & Kratz, K. L. (1998). "Determination of antibiotics in different water compartments via liquid chromatography-electrospray tandem mass spectrometry." *Journal of Chromatography A,* 815: 213–223. doi: 10.1016/S0021-9673(98)00335-5.

[83] Ternes, T. A., Stuuber, J., Herrmann, N., McDowell, D., Ried, A., & Kampmann, M. (2003). "Ozonation: a tool for removal of pharmaceuticals, contrast media, and musk fragrances from wastewater?" *Water Research,* 37: 1976–1982. doi: 10.1016/S0043-1354(02)00570-5.

[84] Berset, J. D., Kupper, T., Etter, R., and Tarradellas, J. (2004). "Consideration about the enantioselectivity transformation of polycyclic musks in wastewater, treated wastewater, and sewage sludge and analysis of their fate in a sequencing batch reactor plant." *Chemosphere,* 57: 987–996. https://doi.org/10.1016/j.chemosphere.2004.07.020.

[85] Watts, C. D., Craythorne, B., Fielding, M., and Killops, S. D. (1982). "Nonvolatile organic compounds in treated waters." *Environmental Health Perspectives,* 46: 87–89. doi: 10.1289/ehp.824687. PMID: 6759110; PMCID: PMC1569043.

[86] Kummerer, K. (2001). "Drugs in the environment: emission of drugs, diagnostic aids, and disinfectants into wastewater by hospitals in relation to other sources—a review." *Chemosphere,* 45: 957–969. http://dx.doi.org/10.1016/S0045-6535(01)00144-8.

[87] Lindberg, R., Jarnheimer, P. A., Olsen, B., Johansson, M., and Tysklind, M. (2004). "Determination of antibiotic substances in hospital sewage water using solid-phase

extraction and liquid chromatography/mass spectrometry and group analogue internal standards." *Chemosphere,* 57: 1479–1488. doi: 10.1016/j.chemosphere. 2004.09.015.

[88] Kinney, C. A., Furlong, E. T., Werner, S. L., and Cahill, J. D. (2006). "Presence and distribution of wastewater-derived pharmaceuticals in soil irrigated with reclaimed water." *Environmental Toxicology and Chemistry,* 25: 317–326. doi: 10.1897/05-187R.1.

[89] Batt, A. L., Bruce, I. B., and Aga, D. S. (2006). "Evaluating the vulnerability of surface waters to antibiotic contamination from varying wastewater treatment plant discharges." *Environmental Pollution,* 142: 295–302. doi: 10.1016/j.envpol.2005.10.010.

[90] Lindsey, M. E., Meyer, T. M., and Thurman, E. M. (2001). "Analysis of trace levels of sulfonamide and tetracycline antimicrobials in groundwater and surface water using solid-phase extraction and liquid chromatography/mass spectrometry." *Analytical Chemistry,* 73: 640–646. doi: 10.1021/ac010514w.

[91] Zuccato, E., Castiglioni, S., and Fanelli, R. (2005). "Identification of the pharmaceuticals for human use contaminating the Italian aquatic environment." *Journal of Hazardous Materials,* 122: 205–209. doi: 10.1016/j.jhazmat.2005. 03.001.

[92] Mutiyar, P.K., & Mittal, A.K. (2013). "Occurrences and fate of an antibiotic amoxicillin in extended aeration-based sewage treatment plant in Delhi, India: a case study of emerging pollutant." *Desalination and Water Treatment,* 51: 31-33: 6158-6164. doi: 10.1080/19443994.2013.770199.

[93] Larsson, D. J. G., de-Pedro, C., & Paxeus, N. (2007). "Effluent from drug manufactures contains extremely high levels of pharmaceuticals." *Journal of Hazardous Materials,* 148: 751–755. doi: 10.1016/j.jhazmat.2007.07.008.

[94] Ramaswamy, B. R., Shanmugam, G., Velu, G., Rengarajan, B., and Larsson, D. G. J. (2011). "GC-MS analysis and ecotoxicological risk assessment of triclosan, carbamazepine and parabens in Indian rivers." *Journal of Hazardous Materials,* 186: 1586–1593. doi: 10.1016/j.jhazmat.2010.12.037.

[95] Kim, Y., Jung, J., Oh, S. R., and Choi, K. (2008). "Aquatic toxicity of cartap and cypermethrin to different life stages of Daphnia magna and Oryziaslatipes." *Journal of Environmental Science and Health. Part. B,* 43: 56–64. doi: 10.1080/03601230701735029.

[96] Oetken, M., Bachmann, J., Schulte-Oehlmann, U., and Oehlmann, J. (2004). "Evidence for endocrine disruption in invertebrates." *International Review of Cytology,* 236: 1–44. doi: 10.1016/S0074-7696(04)36001-8.

[97] Alighardashi, A., Pandolfi, D., Potier, O., and Pons, M. N. (2009). "Acute sensitivity of activated sludge bacteria to erythromycin." *Journal of Hazardous Materials,* 172: 685–692. doi: 10.1016/j.jhazmat.2009.07.051.

[98] DeLorenzo, M. E., and Fleming, J. (2008). "Individual and mixture effects of selected pharmaceuticals and personal care products on the marine phytoplankton species Dunaliella tertiolecta." *Archives of Environmental Contamination and Toxicology,* 54: 203–210. doi: 10.1007/s00244-007-9032-2.

[99] Middleton, J.H., and Salierno, J.D. (2012). "Antibiotic resistance in triclosan tolerant fecal coliforms isolated from surface waters near wastewater treatment plant outflows (Morris County, NJ, USA)." *Ecotoxicology and Environmental Safety,* 1–10. doi: 10.1016/j.ecoenv.2012.10.025.

[100] Shah, S. Q. A., Colquhoun, D. J., Nikuli, H. L., and Sorum, H. (2012). "Prevalence of antibiotic resistance genes in the bacterial flora of integrated fish farming environments of Pakistan and Tanzania." *Environmental Science and Technology,* 46: 8672–8679. https://doi.org/10.1021/es3018607.

[101] Zhang, R., Zhang, G., Zheng, Q., Tang, J., Chen, Y., Xu, W.,Zou, Y., & Chan, X. (2012). "Occurrence and risks of antibiotics in the Laizhou Bay, China: impacts of river discharge." *Ecotoxicology and Environmental Safety,* 80: 208–215. doi: 10.1016/j.ecoenv.2012.03.002.

[102] Desbrow, C., Routledge, E. J., Brighty, G.C., Sumpter, J.P., and Waldock, M. (1998). "Identification of estrogenic chemicals in STW effluent. 1. Chemical fractionation and in vitro biological screening." *Environ. Sci. Technol.,* 32, 11: 1549-1558. https://doi.org/10.1021/es9707973.

[103] Pickering, A.D., Sumpter, J.P. (2003). "Comprehending endocrine disrupters in aquatic environments Environ." *Sci. Technol.,* 37, 17: 331A-336A. doi: 10.1021/es032570f.

[104] Ternes, T.A., Stüber, J., Herrmann, N., McDowell, D., Ried, A., Kampmann, M., and Teiser, B. (2003). "Ozonation: a tool for removal of pharmaceuticals, contrast media and musk fragrances from wastewater?" *Water Res.,* 37: 1976-1982. doi: 10.1016/S0043-1354(02)00570-5.

[105] Burkhardt-Holm, P., Wahli, T., and Meier, W. (2000). "Nonylphenol affects the granulation pattern of epidermal mucous cells in rainbow trout, Oncorhynchus mykiss. Ecotox." *Environ. Safety,* 46: 34-40. doi: 10.1006/eesa.1999.1871.

[106] Burkhardt-Holm, P., Peter, A., and Segner, H. (2002). "Decline of fish catch in Switzerland project fishnet: a balance between analysis and synthesis." *Aquat. Sci.,* 64: 36-54. https://www.academia.edu/15026010/Decline_of_fish_catch_in_Switzerland.

[107] Bielena, A., Šimatovićb, A., Kosić-Vukšićc, J., Senta, I., Ahel, M., Babić, S., Jurina,T., González plaza, J.J., Milaković, M., and Udiković-Kolić, N. (2017). "Negative environmental impacts of antibiotic-contaminated effluents from pharmaceutical industries." *Water Research.* 126: 79-87. doi: 10.1016/j.watres.2017.09.019.

[108] Li, Z., Schulz, L., Ackley, C., & Fenske, N. (2010). "Adsorption of tetracycline on kaolinite with pH-dependent surface charges." *J Colloid Interface Sci.* 351, 1: 254-60. doi: 10.1016/j.jcis.2010.07.034.

[109] Wang, J., and Wang, S. (2016). "Removal of pharmaceuticals and personal care products (PPCPs) from wastewater: a review." *J Environ Manage.* 182: 620-40. doi: 10.1016/j.jenvman.2016.07.049.

[110] Eslami, H., Talebi Hematabadi, P., Ghelmani, S. V., Salehi Vaziri, A., & Derakhshan, Z. (2015). "The Performance of Advanced Sequencing Batch Reactor in Wastewater Treatment Plant to Remove Organic Materials and Linear Alkyl

Benzene Sulfonates." *Jundishapur Journal of Health Sciences.* 7, 3: 33-9. doi: 10.17795/jjhs-29620.

[111] Shahi, D. H., Ebrahimi, A., Esalmi, H., Shirin, A., & Naime, D. (2012). "Efficiency of straw plants in removal of indicator pathogens from sub surface flow constructed wetlands of municipal wastewater in Yazd, Iran" *Journal of Health and Development.* 1, 2: 147-55. https://www.sid.ir/en/Journal/ViewPaper.aspx?ID=299579.

[112] Shahi, D. H., Eslami, H., Ehrampoosh, M. H, Ebrahimi, A., Ghaneian, M. T., Ayotollah, S. & Mozayan, M. R. (2013). "Comparing the efficiency of Cyperus alternifolius and Phragmites australis in municipal wastewater treatment by subsurface constructed wetland." *Pak J. Biol Sci.* 16, 8: 379-84. doi: 10.3923/pjbs.2013.379.384.

[113] Fernandes, J.P., Almeida C.M.R., Pereira A.C., Ribeiro I.L., Reis I., Carvalho P., Basto M.C.P., &Mucha A.P. (2015). "Microbial community dynamics associated with veterinary antibiotics removal in constructed wetlands microcosms." *Bioresour. Technol.*, 182: 26-33. doi: 10.1016/j.biortech.2015.01.096.

[114] Kruglova, A., Kråkström, M., Riska, M., Mikola, A., Rantanen, P., Vahala, R., and Kronberg, L. (2016). "Comparative study of emerging micro-pollutants removal by aerobic activated sludge of large laboratory-scale membrane bioreactors and sequencing batch reactors under low-temperature conditions." *Bioresour. Technol.*, 214: 81-88. doi: 10.1016/j.biortech.2016.04.037.

[115] Krustok, I., Odlare, M., Truu, J., & Nehrenheim, E. (2016). "Inhibition of nitrification in municipal wastewater-treating photobioreactors: Effect on algal growth and nutrient uptake. Bioresour." *Technol.*, 202: 238-243. doi: 10.1016/j.biortech.2015.12.020.

[116] Zhao, X., Chen, Z., Wang, X., Li, J., Shen, J., & Xu, H. (2015). "Remediation of pharmaceuticals and personal care products using an aerobic granular sludge sequencing bioreactor and microbial community profiling using Solexa sequencing technology analysis." *Bioresour. Technol.*, 179: 104-112. doi: 10.1016/j.biortech.2014.12.002.

[117] Vujevic, D., Koprivanac, N., Bozic, L.A., Locke, B., and Locke, B. R. (2004). "The Removal of Direct Orange 39 by Pulsed Corona Discharge from Model Wastewater." *Environ. Technol.* 257: 791–800. doi: 10.1080/09593330.2004.9619370.

[118] Al-kdasi, A., Idris, A., Saed, K., and Guan, C.T. (2004). "Treatment of textile wastewater by advanced oxidation processes—A Review." *Glob. Nest Int. J.* 6: 222–230. https://www.semanticscholar.org/paper/TREATMENT-OF-TEXTILE-WASTEWATER-BY-ADVANCED-A-Al-Kdasi-Idris/917040aa2ae324ce2e434b0147f10aab5f15e72c.

[119] Hoeben, W. F. L. M. (2000). "Pulsed corona-induced degradation of organic materials in water. [Phd Thesis 1 (Research TU/e / Graduation TU/e), Applied Physics]." *Technische Universiteit Eindhoven.* https://doi.org/10.6100/IR535691.

[120] Mouele, E.S.M., Tijani, O.J., Fatoba, O.O., & Petrik, L.F. (2015). "Degradation of organic pollutants and microorganisms from wastewater using different dielectric

barrier discharge configurations—A critical review." *Environ. Sci. Pollut. Res.* 22: 18345–18362. https://doi.org/10.1007/s11356-015-5386-6.

[121] Saracino, M., Emmi, S.S., and Zanelli, A. (2015). "Water remediation 2.0: Advanced Oxidation Processes." *Chimica & Ambiente,* 18–25. doi: http://dx.medra.org/10.17374/CI.2015.97.5.18.

[122] Boczka, G., Fernandes, A., & Makos, P. (2017). "Study of Different Advanced Oxidation Processes for Wastewater Treatment from Petroleum Bitumen Production at Basic pH." *Ind. Eng. Chem. Res.* 56: 8806–8814. https://doi.org/10.1021/acs.iecr.7b01507.

[123] Dewil, R., Mantzavinos, D., Poulios, I., and Rodrigo, A.M. (2017). "New perspectives for Advanced Oxidation Processes." *J. Environ. Manag.* 195: 93–99. doi: 10.1016/j.jenvman.2017.04.010.

[124] Aziz, K.H.H., Miessner, H., Mueller, S., Mahyar, A., Kalass, D., Moeller, D., Khorshid, I., and Rashid, M.A.M. (2018). "Comparative study on 2,4-dichlorophenoxyacetic acid and 2,4-dichlorophenol removal from aqueous solutions via ozonation, photo catalysis and non-thermal plasma using a planar falling film reactor." *J. Hazard. Mater,* 343: 107–115. doi: 10.1016/j.jhazmat.2017.09.025.

[125] Nehra, V., Kumar, A., & Dwivedi, H.K. (2008). "Atmospheric Non-Thermal Plasma Sources." *Int. J. Eng.* 2: 53–68. https://citeseerx.ist.psu.edu/viewdoc/download?doi=10.1.1.517.5732&rep=rep1&type=pdf.

[126] Wardenier, N., Vanraes, P., Nikiforov, A., Van-Hulle, W.H.S., & Leys, C. (2019). "Removal of micro pollutants from water in a continuous-flow electrical discharge reactor." *J. Hazard. Mater.*, 362: 238–245. doi: 10.1016/j.jhazmat.2018.08.095.

[127] Ternes, T.A., Meisenheimer, M., McDowell, D., Sacher F., Brauch, H., Haist-Gulde B., Preuss, G., Wilme, U., & Zulei-Seibert, N. (2002). "Removal of Pharmaceuticals during Drinking Water Treatment." *Environ. Sci. Technol.*36, 17: 3855–3863. doi: 10.1021/es015757k.

[128] Sipma J., Osuna B., Collado N., Monclús H., Ferrero G., Comas J., & Rodriguez-Roda, I. (2010). "Comparison of removal of pharmaceuticals in MBR and activated sludge systems." *Desalination*, 250, 2: 653-659. DOI : 10.1016/j.desal.2009.06.073.

[129] Huang, L., & Lee, D.J. (2015). "Membrane bioreactor: a mini review on recent R&D works." *Bioresour. Technol.*, 194: 383-388. https://doi.org/10.1016/j.biortech.2015.07.013.

[130] Wei, X., Li, B., Zhao, S., Wang, L., Zhang, H., Li, C., & Wang, S. (2012). "Mixed pharmaceutical wastewater treatment by integrated membrane-aerated biofilm reactor (MABR) system – a pilot-scale study. Bioresour." *Technol.*, 122: 189-195. https://doi.org/10.1016/j.biortech.2012.06.041.

[131] De-Gusseme, B., Pycke, B., Hennebel, T., Marcoen, A., Vlaeminck, S.E., Noppe, H., Boon, N., & Verstraete, W. (2009). "Biological removal of 17α-ethinylestradiol by a nitrifier enrichment culture in a membrane bioreactor." *Water Res.*, 43, 9: 2493-2503. https://doi.org/10.1016/j.watres.2009.02.028.

[132] Dawas-Massalha, A., Gur-Reznik, S., Lerman, S., Sabbah, I., & Dosoretz, C. G. (2014). "Co-metabolic oxidation of pharmaceutical compounds by a nitrifying

bacterial enrichment." *Bioresour. Technol.*, 167: 336-342. https://docksci.com/co-metabolic-oxidation-of-pharmaceutical-compounds-by-a-nitrifying-bacterial-enr_5acdde32d64ab29bb15d5551.html.

[133] Snyder, S.A., Adham, S., Redding, A.M., Cannon, F.S., DeCarolis, J., Oppenheimer, J., Wert, E.C., & Yoon, Y. (2007). "Role of membranes and activated carbon in the removal of endocrine disruptors and pharmaceuticals." *Desalination,* 202,1: 156-181. https://doi.org/10.1016/j.desal.2005.12.052.

[134] Kimura, K., Hara, H., & Watanabe, Y. (2005). "Removal of pharmaceutical compounds by submerged membrane bioreactors (MBRs)." *Desalination,* 178, 1:135-140. https://doi.org/10.1016/j.desal.2004.11.033.

[135] Bernhard, M., Müller, J., & Knepper, T. P. (2006). "Biodegradation of persistent polar pollutants in wastewater: comparison of an optimised lab-scale membrane bioreactor and activated sludge treatment." *Water Res.,* 40,18: 3419-3428. https://doi.org/10.1016/j.watres.2006.07.011.

[136] Joss, A., Andersen, H., Ternes, T., Richle, P. R., & Siegrist, H. (2004). "Removal of estrogens in municipal wastewater treatment under aerobic and anaerobic conditions: consequences for plant optimization." *Environ. Sci. Technol.*, 38,11: 3047-3055. doi: 10.1021/es0351488.

[137] Radjenović, J., Petrović, M., & Barceló, D. (2009). "Fate and distribution of pharmaceuticals in wastewater and sewage sludge of the conventional activated sludge (CAS) and advanced membrane bioreactor (MBR) treatment". *Water Res.*, 43,3: 831-841. doi: 10.1016/j.watres.2008.11.043.

[138] Arya, V., Philip, L., & Bhallamudi, M. S. (2016). "Performance of suspended and attached growth bioreactors for the removal of cationic and anionic pharmaceuticals." *Chem. Eng. J.*, 284: 1295-1307. doi: 10.1016/j.cej.2015.09.070.

[139] Thuy, Q. T. T., & Visvanathan, C. (2006). "Removal of inhibitory phenolic compounds by biological activated carbon coupled membrane bioreactor." Water Sci Technol, 53,11: 89-97. doi: 10.2166/wst.2006.341.

[140] Li, X., Hai, F.I., & Nghiem, L.D. (2011). "Simultaneous activated carbon adsorption within a membrane bioreactor for an enhanced micro-pollutant removal. Bioresour." *Technol.*, 102, 9: 5319-5324. doi: 10.1016/j.biortech.2010.11.070.

[141] Serrano, D., Suarez, S., Lema, J.M., & Omil, F. (2011). "Removal of persistent pharmaceutical micropollutants from sewage by addition of PAC in a sequential membrane bioreactor." *Water Res.*, 45,16: 5323-5333. doi: 10.1016/j.watres.2011.07.037.

[142] Golet, E. M., Xifra, I., Siegrist, H., Alder, A. C., & Giger, W. (2003). "Environmental exposure assessment of fluoroquinolone antibacterial agents from sewage to soil." *Environmental Science and Technology*, 37: 3243–3249. doi: 10.1021/es0264448.

[143] McArdell, C., Molnar, E., Suter, M. J. F., & Giger, W. (2003). "Occurrence and fate of macrolide antibiotics in wastewater treatment plants and in the Glatt Valley Watershed Switzerland." *Environmental Science and Technology*, 37: 5479–5486. https://doi.org/10.1021/es034368i.

[144] Lin, A. Y. C., Yu, T. H., & Lateef, S. K. (2009). "Removal of pharmaceuticals in secondary wastewater treatment processes in Taiwan." *Journal of Hazardous Materials*, 167, 1163–1169. doi: 10.1016/j.jhazmat.2009.01.108.

[145] Bendz, D., Paxeus, N. A., Ginn, T. R., and Loge, F. J. (2005). "Occurrence and fate of pharmaceutically active compounds in the environment, a case study: Hoje River in Sweden." *Journal of Hazardous Materials*, 122: 195–204. doi: 10.1016/j.jhazmat.2005.03.012.

Chapter 9

Photocatalytic Mineralization of Pharmaceutical Contaminants in Water by Heterogeneous Photocatalysis in the Presence of LED-Irradiation: A Green and Sustainable Approach

Vijendra Singh Solanki[1,*], Brijesh Pare[2], Neha Agarwal[3] and Har Lal Singh[1]

[1]Department of Chemistry, School of Liberal Arts & Sciences (SLAS), Mody University of Science and Technology, Lachmangarh, Rajasthan, India
[2]Department of Chemistry, Government Madhav Science P.G. College, Ujjain, MP, India
[3]Department of Chemistry, Navyug Kanya Mahavidyalaya, Lucknow, India

Abstract

There is an urgent need to mineralize organic pollutants such as antibiotics and other toxic organic materials discharged by various pharmaceutical and chemical industries into freshwater. These organic pollutants affect the aquatic ecosystem and human health due to bioaccumulation in the food chains and food webs. Heterogeneous photocatalysis is an advanced oxidation-based green technique by which these pharmaceutics can be degraded in the wastewater by using non-toxic and eco-friendly catalysts in the presence of light-emitting diodes (LEDs) as a source of irradiation. The main aim of the chapter is to disseminate information regarding the degradation of antibiotics and other pharmaceutics by green, non-toxic, and effective catalysts via

---

* Corresponding Author's Email: vijendrasingh0018@gmail.com.

In: Pharmaceuticals: Boon or Bane
Editor: Neha Agarwal
ISBN: 979-8-88697-487-4

economically viable techniques. Antibiotics are oxidized by hydroxyl radicals and superoxides generated during the irradiation of light on the surface of the catalyst. In this chapter, the authors discuss the most commonly prescribed antibiotics in the COVID-19 pandemic and how these antibiotics become environmental contaminants. They have also proposed the mechanism of degradation of these antibiotics in the presence of LED irradiation to attain a green and sustainable environment.

Keywords: antibiotics, LED-irradiation, catalyst, green chemistry, AOPs

1. Introduction

The present scenario of the world has totally changed during the COVID-19 pandemic. The extensive use of pharmaceuticals and medical equipment has affected the environment directly or indirectly and disturbed the life cycle of flora and fauna. A large amount of medical and industrial waste is dumped improperly into water bodies and soil, either directly or indirectly. A study revealed the rate of consumption of antibiotics is very high in high-income group countries such as the U.S., Italy, and France, and these countries have been affected more by the COVID-19 pandemic as compared to the low-income group countries. India, China, and Pakistan were the highest consumer countries of antibiotics that were affected the most by COVID-19. On the other hand, a European CDC report based on antibiotic consumption revealed that Europe and other countries like Spain, France, Italy, Ireland, Belgium, Portugal, Poland, and the U.K. were also high antibiotic users – some of which suffered greatly from the COVID-19 pandemic [1]. Antibiotics and other pharmaceuticals are constantly polluting water bodies, and only 1.1 percent of water is potable, with the rest contaminated by pharmaceutical organic molecules. Improper and irrational use of pharmaceuticals affects the natural properties of water, human health, and the environment [2, 3].

In the year 2019, Patel et al. and Berebdonk et al. (2015) reported that antibiotic-resistant bacteria and antibiotics were detected as the most frequent pollutants in water bodies and the environment [4, 5]. Similarly, other pharmaceuticals such as caffeine was detected in waste-water bodies in the United States, estrogens in Germany, theophylline and tetracycline (in 1 μg/L) in U.K., analgesic phenacetin in Spain, ibuprofen and naproxen as sewage effluents in Canada, clofibric acid in Germany, salicylic acid, meprobamate, pentobarbital, phensuximide and sulfonamides in the United States, sulfonamides and barbiturates in Denmark, lipid regulating agents (up to 5 μg/L) and NSAIDs in Germany, and both β-blockers and antibiotics (up to 6 μg/L) in Germany [6-19].

In 2019, the U.S. Center for Disease Prevention and Control (2019) reported that 50 percent of the antibiotics are not medically justified, and developing countries are greatly affected by them since they do not have proper system to degrade such types of antibiotics. It is observed by the "UN Interagency Coordination Group on Antimicrobial Resistance" that around 700,000 people die annually. It is estimated that this ratio will reach up to 10 million per year by 2050 [20, 21, 22].

Hence, it is necessary to mineralize these antibiotics into an aqueous medium. Physicochemical processes such as reverse osmosis, ultrafiltration, and coagulation have been used to degrade a class of pharmaceuticals, but they require expensive chemical treatment plants, high operational costs, and these processes produce massive amounts of secondary pollutants, such as sludge. Moreover, some pharmaceutics have characteristic properties such as lipophilicity, polarity, volatility, adsorption properties, and persistence. As a result, simple and conventional waste water treatment technologies can not remove these pharmaceuticals form the aqueous environment [23, 24]. Recently, heterogeneous photocatalysis is considered to be a suitable, efficient, cost-effective and green technique by which various classes of antibiotics and other pharmaceutical compounds can be degraded successfully. Metal oxides, halides, mixed metal composites, doped and co-doped metal oxides are used to degrade antibiotics in an aqueous medium by using eco-friendly LED irradiation and UV-Visible light sources [25, 26].

Mineralization of antibiotics and other organic pollutants is based on the production of principal hydroxyl radicals ($^{\bullet}OH$) and other oxidizing radicals such as hydroperoxyl ($HO_2^{\bullet}$) and superoxide ($O_2^{-\bullet}$) during the irradiation of light on the surface of the photocatalyst in an aqueous medium. These oxidizing ions degrade antibiotics and other pharmaceuticals into CO_2 and H_2O [27, 28, 29].

2. Occurrence of Pharmaceuticals in the Environment

The continuous use of antibiotics and their production by pharmaceutical industry can adversely affect natural environmental resources such as water, soil, and vegetables. Regular and uncontrolled usage can develop resistance to antibiotics. Hence, major diseases cannot be treated by similar kinds of antibiotics. It is reported that in the European Union, around 33,000 people and in the United States, around 35,000 people die every year, while 2.8 million are affected by antibiotic resistant bacteria. Furthermore, antibiotic-resistant bacteria are more dangerous for humans [30]. Statistical data shows that the use of antibiotics in animal husbandry is much higher than in the treatment of people. According to Van Boeckel et al.'s first global (228 countries) map of antibiotic consumption in livestock, total consumption of antibiotics in 2010 was 63 percent and expected to rise by 67 percent by 2030, while it would have doubled in BRICS countries such as Brazil, Russia, India, China, and South Africa by that time [31, 32, 33].

3. Frequently Prescribed Antibiotics during COVID 19 Pandemic

Azithromycin (Zithromax) is one of the important antibiotics which are prescribed by doctors in the treatment of COVID-diseases. It was often used in combination with hydroxychloroquine. The use of hydroxychloroquine is not recommended for the treatment of COVID-19 by the National Institute of Health (NIH) [34].

Figure 1 shows the percentage of antibiotics that were used in the COVID-19 pandemic, and Figure 2 shows the administration of antimicrobials during COVID-19[35].

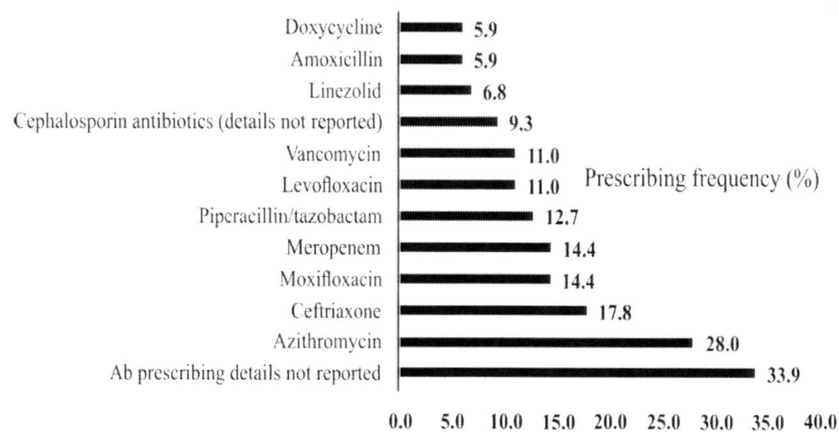

Figure 1. Prescribed percentage of antibiotics during COVID 19.

Figure 2 shows the administration of antimicrobials during the COVID-19 pandemic in different hospitals and the implications for antimicrobial resistance in health sectors.

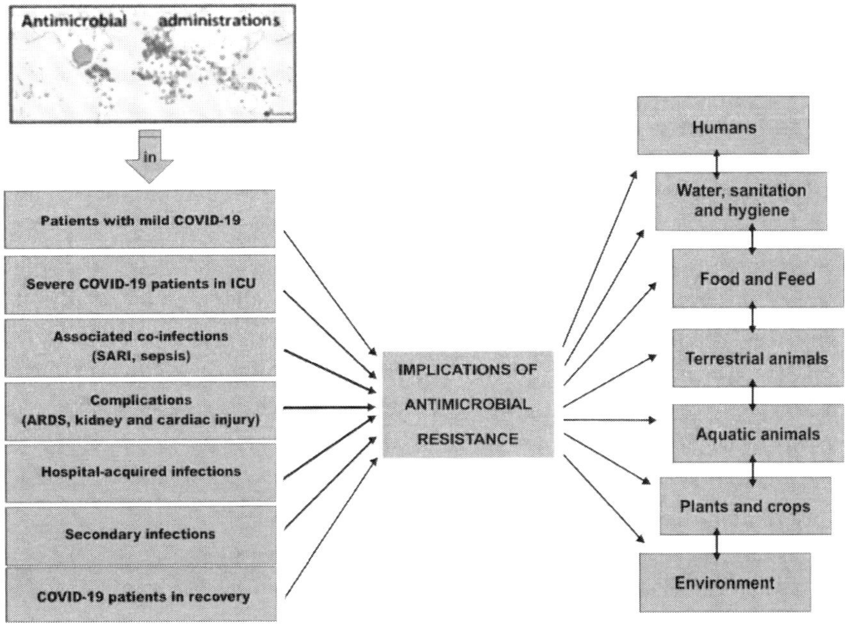

Figure 2. Administration of antimicrobials and the implications for antimicrobial resistance in health sectors [36].

3.1. Some Important Antibiotics

Antibiotics and raw materials that are used in the synthesis of antibiotics by pharmaceutical and medicinal companies are considered potential hazardous substances since their continuous exposure into the environment and atmosphere affects the flora and fauna. Hence, scientists and researchers have another interest in solving this serious problem and protecting human health and environment from these toxic materials. Extensive and continuous consumption of antibiotics by humans is not only dangerous for the environment but affects aquatic life directly or indirectly [37]. Many research papers and reports have been published on the causes and problems created by antibiotics, but there is no concrete idea to solve this serious problem and protect the environment from the huge accumulation of antibiotics [38].

Researchers have worked on the degradation of penicillin in aqueous medium, but fluoroquinolones and tetracyclines are another class of antibiotics that are not degraded easily by simple and eco-friendly methods. These antibiotics affect the environment and water due to their long-time availability in the water and soil. A study revealed that about 0.9 mg of ciprofloxacin per gram of organic compound was found in river sediments in Patancheru, near Hyderabad (India), released from industrial power plants. As a result of the waste generated by pharmaceutical industries, most of the region has been polluted by these antibiotics and has totally changed the environment of that region. For example, a particular area of Hyderabad has been polluted since regulations related to the release of pollutants and active ingredients into the environment are not obeyed by various countries like India, China, Europe, and the U.S. [39]. Some important categories of antibiotics are described below.

3.1.1. Penicillins

According to the molecular structure reported in Figure 3, penicillins belong to beta-lactum class of antibiotics. This class of antibiotics contains five groups of antibiotics:

a) aminopenicillins,
b) antipseudomonal penicillins,
c) beta-lactamase inhibitors,
d) natural penicillins,
e) penicillinase resistant penicillins.

These are used in various types of bacterial infections.

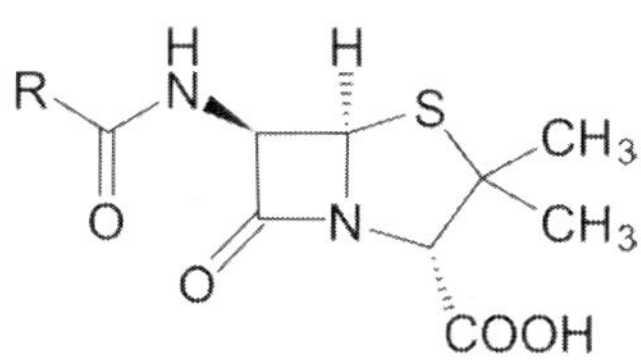

Figure 3. Penicillin.

3.1.2. Tetracyclines

This is broad-spectrum class of antibiotic which is generally prescribed by doctors for the treatment of following infections:

a) acne related infections,
b) urinary tract infections (UTIs),
c) intestinal tract infections,
d) eye infections,
e) sexually transmitted diseases, and
f) periodontitis (gum disease)

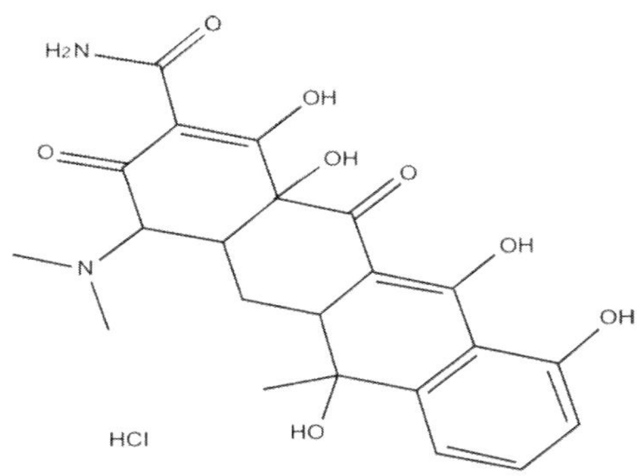

Figure 4. Tetracyclin.

3.1.3. Cephalosporins

The cephalosporins are generally used for the treatment of animals and humans who are infected by gram-negative bacteria. Its action is similar to penicillin and is used to kill bacteria. The structure is mentioned in Figure 5.

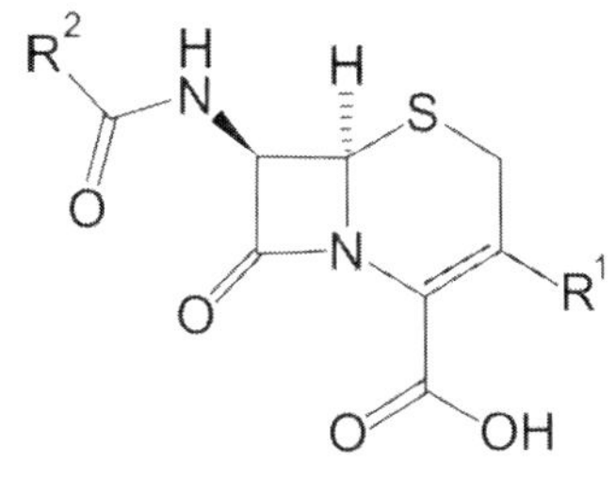

Figure 5. Cephalosporins.

3.1.4. Fluoroquinolones

Fluoroquinolones belong to the class of broad spectrum antibiotics that are generally used for adults but cannot be used for the treatment of children. These are also known as quinolones, which are synthetic, broad-spectrum bactericidal and antibacterial classes of antibiotics. Since it generates side effects in the body, the Food & Drug Administration (FDA) has advised and suggested to doctors that they are not suitable for the treatment of various general diseases such as: (a) sinusitis, (b) bronchitis, and (c) uncomplicated urinary tract problems and infections.

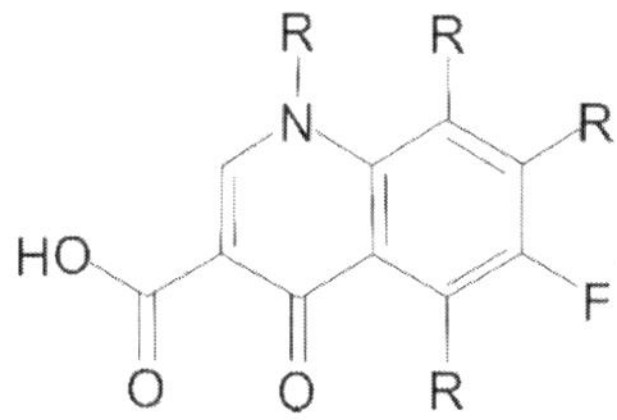

Figure 6. Fluoroquinolones.

3.1.5. Lincomycins

Lincomycins are used to treat diseases which are caused by different types of gram-positive aerobic and gram-negative anaerobic bacteria. Generally, it is used in the treatment of pelvic inflammatory problems and infections, intra-abdominal problems, lower respiratory tract related infections and diseases, bones and joint related problems, and is occasionally used in the treatment of acne. These are also used in the treatment of vaginosis infections caused by bacteria, Figure 7.

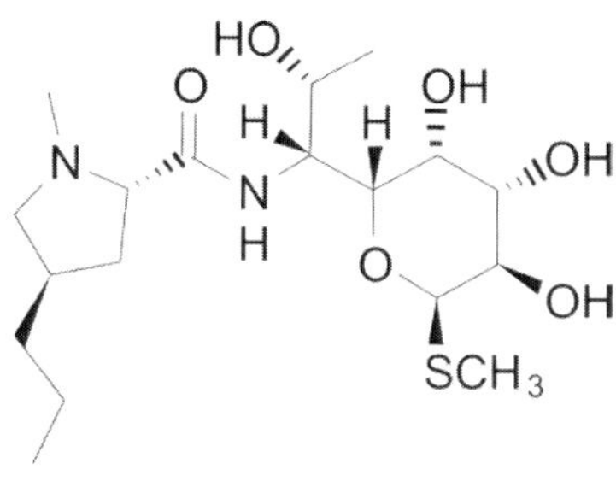

Figure 7. Lincomycin.

3.1.6. Macrolides

This is another important class of antibiotics that are used in the treatment of pneumonia, pertussis (also known as whooping cough), skin problems, and other susceptible problems. The general structure of macrolide is mentioned in Figure 8.

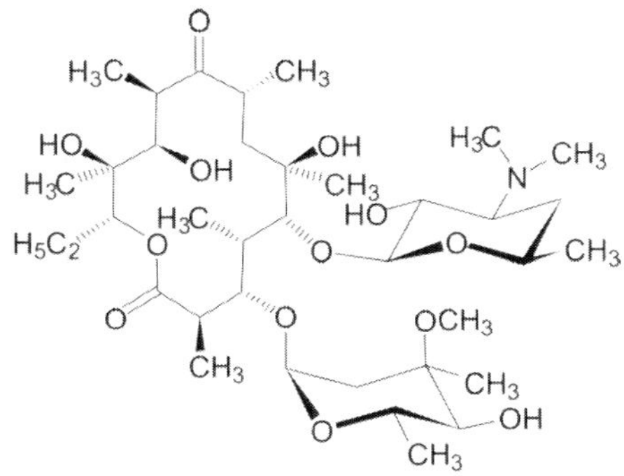

Figure 8. Macrolide.

3.1.7. Glycopeptide Antibiotics

These are used in the treatment of methicillin-resistant *staphylococcus aureus* (MRSA) infections, complicated skin infections, *C. difficile*-associated diarrhea, and enterococcal infections such as endocarditis that are resistant to beta-lactams and other antibiotics.

3.1.8. Aminoglycosides

These types of antibiotics are used to treat and reduce bacterial synthesis by binding to the 30 S ribosome and act as bactericidal antibiotics. Doctors typically prescribe these medicines to the patients who have serious eye infections.

3.1.9. Carbapenems

These belong to the beta-lactam class and are capable of killing various types of bacteria. They are often used for serious infections or as "last-line" agents for stomach infections, pneumonia, kidney infections, and multidrug-resistant hospital-acquired infections. The structure of carbapenemis mentioned in Figure 9.

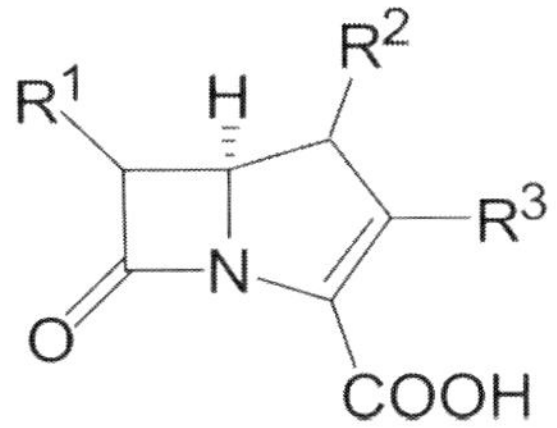

Figure 9. Carbapenem.

3.1.10. CeftriaxoneSodium

This is another important third-generation cephalosporin class of antibiotic, which is used in the treatment of bacterial infections of the human body and is used by medical practitioners and doctors [40].

3.1.11. Sulfamethoxazole

Sulfamethoxazole is a sulfonamide-based antibiotic that is used for bacterial infections. The previously published data showed that (https://toxics.usgs.gov/highlights/antibiotics_gw/background.html) it is a contaminant in U.S. streams and groundwater where wastewater treatment plants and septic tanks are the sources of the release of antibiotics to the environment.

4. Environmental Contamination by Antibiotics

Various pharmaceutical industries manufacture antibiotics and doctors prescribe them to treat different types of infections in human and animal bodies. These antibiotics are disposed of in the form of unused raw and waste materials and become a source of environmental contamination. The consumed antibiotics are not completely metabolized in the body; hence, the remaining portion is exerted by urine or soot into the environment. Industrial processes like the production of ethanol and fertilizers, when used by farmers and others, are discharged into fresh water bodies and contaminate the environment. The discharged antibiotics in water and other bodies are consumed by bacteria (which are present in the soil and water).Hence, they develop resistance to a few antibiotics and are not affected next time. Since bacteria change genetically according to their class itself, hence showing resistance to antibiotics.

5. How to Save the Environment from the Antibiotics

As we know that the environment is greatly affected by the release and accumulation of different categories of xenobiotics like pharmaceutical by-products, personal care products (PCPs), and fertilizers, there is a need to save the environment from the toxic effects of these pollutants [41]. Following are the important preventive action plans, by implementing them, the pollution could be controlled.

- Use antibiotics in that condition when it's urgent, otherwise avoid their use. Wash your hands properly after using them and don't throw them anywhere.
- Do not ask for antibiotics if your medical consultant or veterinarian does not recommend them.
- Use only prescribed antibiotics directed by physicians or doctors.
- Do not keep antibiotics for future use since they create adverse effects on the body even when kept in the environment [42].

6. Photocatalysts for the Degradation of Pollutants

A variety of metal oxide photocatalysts such as titanium dioxide (TiO_2), zinc oxide (ZnO), tungsten oxide (WO_3), strontium titanate ($SrTiO_3$), and Cerium oxide (CeO_2) work as unique catalysts which have been synthesized and categorized into metal/CeO_2, metal oxide/CeO_2, CeO_2/support, such as Co/CeO_2, Ni/CeO_2, AuCu/CeO_2, MnO_2/CeO_2, CuO/ CeO_2, CeO_2/SiO_2, CeO_2/g-C_3N_4, CeO_2/Al_2O_3, Ni–La_2O_3–CeO_2/SBA, Ni–CeO_2/graphene metal chalcogenides ZnS, CdS, CdSe, WS_2, and hematite (Fe_2O_3), Ag/CeO_2[43, 44, 45, 46, 47, 48, 49, 50]. Similarly, Ag-based catalyst such as Ag/TiO_2, Ag/SBA, Ag/Al_2O_3, Ag/g-C_3N_4 [34, 35] and Ag/CeO_2 catalysts have also been prepared and used in the degradation of various organic pollutants [51, 52, 53, 54].

Most of the catalysts (except for hematite) are effective in the UV region since their bandgap energy is more than or equal to 3.23eV. These catalysts absorb only 7% of UV light from the sun's spectra and most of them are absorbed by the ozone layer, so they are not effective in the degradation of antibiotics and other pharmaceuticals in the visible region of electromagnetic radiation. As it has already been reported and accepted by researchers and scientists that ZnO and TiO_2 are the most powerful and effective catalysts in UV light for the removal of various organic, pharmaceuticals, and toxic pollutants in the aqueous environment. Therefore, the development and synthesis of new visible light active photocatalytic materials are required for popularization of heterogeneous advanced oxidation techniques [55].

Photocatalysts which are being used in the various photocatalytic systems are summarized in Table 1.

Table 1. Photocatalysts with different energy band gap

Material	Band gap (eV)	Wavelength corresponding to band gap (nm)
SnO_2	3.9	318
ZnS	3.7	336
$SrTiO_3$	3.4	365
$BaTiO_3$	3.3	375
ZnO	3.2	390
TiO_2	3.0	390
WO_3	2.8	443
$BaCrO_4$	2.6	465
CdS	2.5	497
GaP	2.3	540
Fe_2O_3	2.2	565
CdO	2.1	590
CdSe	1.7	730
GaAs	1.4	887

The right column in the table shows the amount of radiation required to activate the photocatalyst. Plank has described and given a mathematical equation related to energy and band-gap energy.

$$E = hc/E_G$$

Where 'E_G' is the catalyst band-gap energy, 'h' is Planck's constant, and 'c' is the velocity of light.

A good photocatalyst should have the following properties:

- It should be inexpensive and non-toxic
- Having a high surface area
- Having a broad absorption spectra with high absorption coefficients
- Exhibits tunable properties, which can be modified by size reduction, doping, and sensitizers
- Can be afforded for the multi-electron transfer process
- Capable of extended use without substantial loss of photocatalytic activity.
- Semiconductor particles recovered by filtration or centrifugation or when immobilized retain much of their native activity after repeated catalytic cycles.
- The light absorbing property of the photocatalyst
- Recombination rate of agitated electrons and holes

- Rate for oxidative and reductive reactions that transform electrons and holes to free radicals.

7. Photocatalysis and Advanced Oxidation Processes (AOPs)

Photocatalysis is an eco-friendly and cost-effective technique in which organic molecules are degraded by a variety of catalysts in the presence of various sources of light irradiation. When water is contaminated by non-biodegradable organic pollutants, AOPs can be considered safe and sustainable for the purification of industrial wastewater. Various classes of organic pollutants can be oxidized by AOPs in an aqueous medium. Various types of toxic organic materials, such as chlorinated pesticides, PCPs, organic dyes, plasticizers, microcystin, and toxic metal ions, are effectively mineralized into CO_2 and H_2O using various AOPs. Among the AOPs, heterogeneous AOPs have taken great consideration since they are one of the most powerful, safe, and cost-effective green techniques as compared to the other traditional purification techniques. The heterogeneous photocatalytic based AOPs have many important and useful benefits, such as their establishment and operation being simple and low cost, and LED lights can be used as a sustainable and beneficial source of irradiation [56].

7.1. Non-Photochemical AOPs

In these processes, hydroxyl radicals are generated by ozone/H_2O_2, ozone/catalyst, and ozone/Fentons reagent in the absence of light. The following AOPs take place in the absence of light.

- Ozone based AOPs
  O_3/H_2O_2: Peroxone is a combination of O_3 and H_2O_2. In this method, hydroxyl radicals are formed by the conjugated base (HO_2^-) of hydrogen peroxide [57].
  $H_2O_2 \rightarrow HO_2^- + H^+$
  $HO_2^- + O_3 \rightarrow HO^{\cdot}_2 + O^{3\cdot-}$
  $2O_3 + H_2O_2 \rightarrow 2^{\cdot}OH + O^{3\cdot-}$Ozone + catalyst ($O_3$/CAT)
- Electrochemical oxidation

- Radiolysis and electron-beam treatment
- Non-thermal plasma (surface corona discharge)
- Electro hydraulic discharge-ultrasound (US, cavitations)
- Wet air oxidation
- Supercritical water oxidation
- Zerovalent iron (ZVI)
- Ferrate
- Fenton system (H_2O_2/Fe^{2+})

7.2. Photochemical AOPs

7.2.1. O_3/UV

It is reported that the value of molar extinction coefficient for ozone is 3300 M^{-1} cm^{-1} and it strongly absorbs UV light of wavelength 254 nm. For this reason, ozone is converted into hydrogen peroxide by absorbing UV light from the mercury lamp (as a source of energy) and produces more hydroxyl radicals in the aqueous medium. As a result, most organic compounds and pharmaceuticals can be degraded into carbon dioxide and water by AOPs [58].

$$O_3 + H_2O \rightarrow H_2O_2 + O_2$$
$$H_2O_2 \rightarrow 2{\cdot}OH$$

7.2.2. H_2O_2/UV

The combination of hydrogen peroxide with UV radiation form ·OH radicals by homolytic cleavage. The efficiency of photolysis of hydrogen peroxide is dependent on the absorption of UV energy, which ultimately depends on the molar extinction coefficient (19.6 M^{-1} cm^{-1}) [59].

$$H_2O_2 \rightarrow 2{\cdot}OH$$
$$H_2O_2 + {\cdot}OH \rightarrow H_2O + HO{\cdot}_2$$
$$H_2O_2 + HO{\cdot}_2 \rightarrow H_2O + O_2 + {\cdot}OH$$

7.2.3. O_3/ H_2O_2/UV

Combination of the above two.

7.3. Photo-Fenton/Fenton-Like Systems

This is another type of AOP, which is the combined form of the Fe^{+2}/Fe^{+3} and H_2O_2. In this processes, hydroxyl radicals (·OH) are formed by photo-reduction of Fe^{+3} to Fe^{+2}. The Fe^{+2} again react with H_2O_2 [60].

$$Fe^{+3} + H_2O \rightarrow Fe^{+2} + H^+ + \cdot OH$$

In acidic medium (at pH=3), $Fe(OH)_2^+$ complex is formed which under UV radiation generates Fe^{+2} and ·OH.

$$Fe(OH)_2^+ \rightarrow Fe^{+2} + H^+ + \cdot OH$$

Various photocatalytic systems generate hydroxyl radicals by different ways-

$$H_2O + O_3 \rightarrow HO_3^+ + OH^- \rightarrow 2HO_2 + \cdot OH + 2O_2$$
$$H_2O_2 + h\nu \rightarrow 2\ HO\cdot$$
$$Fe^{2+} + H_2O_2 \rightarrow Fe(OH)^{2+} + HO\cdot$$
$$Fe^{3+} + H_2O_2 \rightarrow Fe^{2+} + HO_2\cdot + H^+$$
$$O_2 + H_2O \rightarrow O_2^{\cdot -} + HO\cdot$$

AOPs are unique and have great advantages because they degrade antibiotic and medicinal compounds (present in the wastewater) into CO_2 and H_2O with the ·OH radical (standard redox potential of 2.8 eV) generated by heterogeneous photocatalysis. The main mechanism is the mineralization of antibiotics and other pharmaceuticals by $OH^{\bullet}$, a process that occurs mainly on the surface of catalysts where $^{\bullet}OH$ radicals are formed. The second mechanism that can lead to oxidation is the possible excitation of antibiotics and other pharmaceuticals with the help of a photocatalyst surface electron-hole and subsequent oxidation.

A proposed mechanism of degradation of antibiotics and pharmaceuticals by heterogeneous photocatalysis is:

$$\text{Photocatalyst} + h\nu_{visible} \rightarrow \text{Photocatalyst}\ (h^+_{VB} + e^-_{CB})$$
$$\text{Photocatalyst}\ (h^+_{VB}) + H_2O \rightarrow \text{Photocatalyst} + H^+ + HO$$
$$\text{Photocatalyst}\ (h^+_{VB}) + OH^- \rightarrow \text{Photocatalyst} + HO^{\bullet}$$
$$\text{Photocatalyst}\ (e^-_{CB}) + O_2 \rightarrow \text{Photocatalyst} + O_2^{\cdot -}$$

$O_2^{\cdot-} + H^+ \rightarrow HO^{\cdot}_2$

$HO^{\cdot}_2 + H^+ + \text{Photocatalyst } (e^{-}_{CB}) \rightarrow H_2O_2 + \text{Photocatalyst}$

$H_2O_2 + \text{ Photocatalyst } (e^{-}_{CB}) \rightarrow HO^{\cdot} + OH^- + \text{ Photocatalyst}$

Antibiotics or other Pharmaceuticals $+HO^{\cdot} \rightarrow$ Degradation Products

Antibiotics or other Pharmaceuticals $(h^+_{VB}) \rightarrow$ Oxidation Products

Antibiotics or other Pharmaceuticals + $(e^-_{CB}) \rightarrow$ Reduction Products

Significant degradation of various pharmaceutics and antibiotics was observed by coupling of O_3 with UV. In the UV/O_3 method, ozone reacts with UV light to form a very reactive oxygen atom (^1D) and O_2, and resulted organic pollutants can be degraded easily and effectively [61, 62, 63, 64].

7.4. Advantages of AOPs

- Inherent destructive nature
- No mass transfer involved
- Can be carried out under ambient conditions and using atmospheric oxygen as the oxidant.
- May lead to complete mineralization of organic carbon into CO_2.

8. Principles of Photocatalysis and Its Types

Various antibiotics and pharmaceutical compounds are degraded by photocatalysis and have gained much interest due to energy and environmental issues. The term "photocatalysis" comprises light and catalysts that are important in photocatalytic reactions. These reactions were carried out by the use of catalysts and light. In a heterogeneous system, antibiotics are degraded in the presence of solid catalysts and different light sources, whereas in a homogeneous system a single-phase system is used. There are some basic features of these reactions that favor their applicability in waste water treatment technologies.

- The process takes place at an ambient temperature.
- Oxidation of the substances into CO_2 is complete.
- The oxygen necessary for the reaction is obtained from the atmosphere.

- The catalyst is cheap, innocuous, and can be reused.
- The catalyst can be attached to different types of inert matrices[12].

8.1. Advantages of LEDs Light

- It has a long-life period compared to other sources like halogens and CFLs.
- Green and environmentally friendly light sources.
- It doesn't create soil and water pollution.
- LEDs use less energy than any other sources of light; hence consumption of energy can be reduced.
- LED lights are up to 80% more efficient, effective and sustainable than conventional sources of light. LEDs convert 95% of energy into lights while fluorescent lights are capable of converting only 5% into light and 95% into heat. LED lights also reduce greenhouse gases by consuming less power as compared to traditional or conventional sources of light.
- LED lights contain nontoxic elements and help to protect the environment from further toxic waste. While the filament of a traditional halogen lamp is made up of tungsten and mercury, this will contaminate the environment when discarded in soil and water.
- LEDs have been designed in such a manner that no extra energy is transmitted hence reducing energy consumption and providing benefits to the environment.
- LEDS have a six-time longer life span as compared to other sources of light. It contributes to pollution reduction by emitting a low carbon percentage of carbon into the environment [61, 65].

9. Move towards Green Chemistry

The concept of green chemistry was introduced by an American chemist in 1996, and the word "green" was adopted in 1996. Further In 1998, Professor Paul Anastas formulated 12 principles of green chemistry. The main aim of green chemistry is to design safer techniques and environmentally friendly compounds that reduce or eliminate the use and generation of hazardous substances in the environment. In the year 1991, (https://www.britannica.com/

topic/Environmental-Protection-Agency) the U.S. Environmental Protection Agency announced another pollution prevention program. The main aim of this programme was to reduce or prevent the environment from toxic pollutants and should be adopted by industries during the synthesis and discharging of these toxic pollutants into the environment. This research program was later expanded to include the development of greener, safer, and eco-friendly chemicals. Finally, the term "*green chemistry*" was adopted in 1996.The following 12 principles of green chemistry were adopted in 1998:

1. *Prevention:* Prevention is always better than cure; we must prevent the release of contaminants into the environment or should clean up the waste that has been created.
2. *Atom Economy:* Those synthetic methods should be designed that can maximize the incorporation of all used materials into the final product.
3. *Less hazardous chemical Synthesis:* Wherever practicable, synthetic methods should be designed to generate those substances that have little or no toxicity on human health and the environment.
4. *Design Safer Chemicals:* Chemical products should be designed to create their desired function with minimum toxicity.
5. *Safer Solvents and Auxiliaries:* The use of auxiliary substances (e.g., solvents, separation agents, etc.) should be made unnecessary, wherever possible and innocuous when used.
6. *Design for energy efficiency:* The energy requirements of different chemical processes must be recognized so that their environmental and economic impacts can be minimized. If possible, synthetic methods should be conducted at ambient temperature and pressure.
7. *Use of renewable feedstock:* Whenever it is technically and economically practicable, a raw material or feedstock should be made renewable rather than depleting.
8. *Reduce derivatives:* Unnecessary derivatization (use of blocking groups, protection/deprotection, and temporary modification of physical/chemical processes) should be minimized or avoided because these steps require additional reagents which in turn can generate waste.
9. *Catalysis:* Catalytic reagents should be used because they are superior to stoichiometric reagents.

10. *Design for degradation:* Chemical products should be designed so that at the end of their function, they break down into innocuous degradation products and do not persist in the environment.
11. *Real-time analysis for pollution prevention:* Analytical methodologies need to be developed further to allow for real-time, in-process monitoring and control prior to the formation of hazardous substances.
12. *Inherently safer chemistry for accident prevention:* The substances used in a chemical process should be chosen in such a manner that minimizes the potential for chemical accidents, including releases, explosions and fires.

10. Mechanism of Degradation of Antibiotics and Other Organic Pollutants

The highly reactive oxygen species and especially hydroxyl radical ($^{\bullet}OH$) play a key role in the mineralization of antibiotics and other organic pollutants in aqueous medium. This method has been successfully implemented for the degradation of different xenobiotics, such as pharmaceuticals, organic dyes, pesticides, and many other toxic and non-biodegradable pollutants. The overall degradation of the target organic pollutant is confirmed by the production of carbon dioxide and water in the aqueous system. The mechanism of heterogeneous photocatalysis is is very complex in nature. Antibiotics and organic pollutants interact with O_2^-, $\cdot OH_2$, or OH^- species to generate intermediates that ultimately lead to the formation of degradation products. Hydroxyl radical ($\cdot OH$) being a very strong oxidizing agent (2.8 eV) mineralizes dye to end product [66, 67]. The mechanism of degradation is shown in Figure 10 and mechanism of transformation pathways of organic pollutants is shown as follows.

Antibiotics/organic pollutants +$HO^{\bullet}$→Decolorization of Antibiotics organic pollutants

↓

Degradation of Antibiotics/
organic pollutants involving organic intermediates

↓

Mineralization

↓

e^{-} + Pharmaceuticals → Pharmaceuticals−
(Direct reduction of pharmaceuticals)
↓
h^{+} + Pharmaceuticals → Pharmaceuticals+
(Direct oxidation of pharmaceuticals)
↓
CO_2 (gas) + H_2O + Inorganic ions according to structure of Antibiotics/organicPollutants.

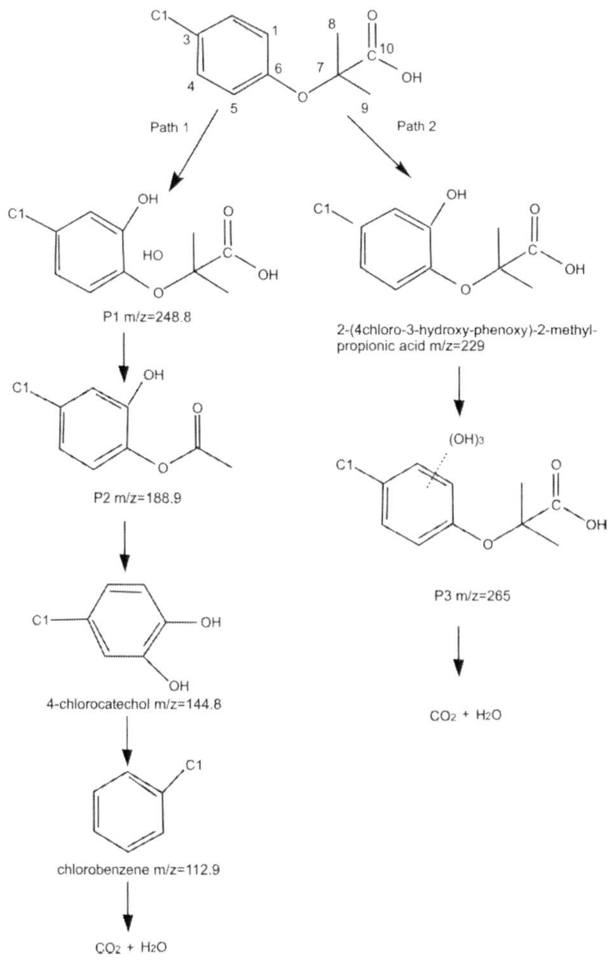

Figure 10. Pharmaceuticals transformation pathway in AOPs bio-electro-Fenton (BEF) system Adopted from [68].

Conclusion

Various organic pollutants, antibiotics, and their toxic effects have been discussed in this chapter. It is concluded that LED irradiation, as a source of visible light is very effective for the degradation of these pollutants in the presence of different visible light-responsive nanomaterials. This technique obeys green chemistry principles and favours the sustainability of heterogeneous photocatalysis. Heterogeneous photocatalysis is a suitable and eco-friendly technique in waste water treatment technology. The discussed mechanism shows that antibiotics and other organic molecules are degraded in CO_2 and H_2O without producing any hazardous materials during the mineralization of the antibiotics and other pollutants.

References

[1] Muhammad, A. (2019). "COVID-19 Pandemic - Link to History of Antibiotic Consumption and Environmental Contamination." https://iep.berkeley.edu/content/covid-19-pandemic-link-history-antibiotic-consumption-and-environmental-contamination.

[2] Ignasi S., & https://www.sciencedirect.com/science/article/pii/S0160412011002108. Enric, B. (2012). "Remediation of water pollution caused by pharmaceutical residues based on electrochemical separation and degradation technologies: a review." *Environ Int.*, 40: 212– 229. https://doi.org/10.1016/j.envint.2011.07.012.

[3] Rivera-Utrilla J., Sánchez-Polo M., Ferro-García M. Á., Prados-Joya G., & Ocampo-Pérez, R. 2013. "Pharmaceuticals as emerging contaminants and their removal from water. A review." *Chemosphere.* 93: 1268–1287. DOI: 10.1016/j.chemosphere.2013.07.059.

[4] Patel, M., Kumar, R., Kishor, K., Mlsna, T., Pittman, C.U., and Mohan, D. (2019). "Pharmaceuticals of emerging concern in aquatic systems: chemistry, occurrence, effects, and removal methods." *Chem. Rev.,* 119: 3510–3673. https://doi.org/10.1021/acs.chemrev.8b00299.

[5] Berendonk, T. U., Manaia, C. M., Merlin, C., Fatta-Kassinos, D., Cytryn, E., Walsh, F., Bürgmann, H., Sørum, H., Norström, M., Pons, M.-N., Kreuzinger, N., Huovinen, P., Stefani, S., Schwartz, T., Kisand, V., Baquero, F., & Martinez, J.L. (2015). "Tackling antibiotic resistance: the environmental framework." *Nat. Rev. Microbiol.* 13: 310–317. doi: 10.1038/nrmicro3439.

[6] Sievers, R. E., Barkley, R. M., Eiceman, G. A., Shapiro, R. H., Walton, H. F., Kolonko, K. J., & Field, L. R. (1977). "Environmental Trace Analysis of Organics in Water by Glass Capillary Column Chromatography and Ancillary Techniques: Products of Ozonolysis." *J. Chromatogr. A.*, 142: 745−754. doi: 10.1016/S0021-9673(01)92082-5.

[7] Sheldon, L. S., & Hites, R. A. (1978). "Organic Compounds in the Delaware River." *Environ. Sci. Technol,* 12, 10: 1188–1194. doi: 10.1021/es60146a006.

[8] Tabak, H. H., & Bunch, R. (1970). "Steroid Hormones as Water Pollutants part I. Metabolism of Natural and Synthetic Ovulation Inhibiting Hormones by Microorganisms of Activated Sludge and Primary Settled Sewage." In *Dev. Ind. Microbiol.*Edited by Corum, C.J. 11: 367–376. American Institute Of Biological Sciences: Washington, DC, USA https://eurekamag.com/research/027/441/027441923.php.

[9] Bunch, R. L. (1981). "Steroid Hormones as Water Pollutants. II. Studies on the Persistence and Stability of Natural Urinary and Synthetic Ovulation-Inhibiting Hormones in Untreated and Treated Wastewaters." In *Dev. Ind. Microbiol., 520.* https://eurekamag.com/research/029/212/029212350.php.

[10] Aherne, G. W., & Briggs, R. (1989). "The Relevance of the Presence of Certain Synthetic Steroids in the Aquatic Environment." *J. Pharm. Pharmacol.,* 41: 735–736. https://doi.org/10.1111/j.2042-7158.1989.tb06355.x.

[11] Shore, L. S., Gurevitz, M., & Shemesh, M. (1993). "Estrogen as an Environmental Pollutant. Bull. Environ." *Contam. Toxicol.,* 51, 3: 361–366. doi: 10.1007/BF00201753.

[12] Waggott, A. (1981). "Trace Organic Substances in the River Lee. In The Chemistry of Water Reuse; Cooper." *Chemical Reviews Review*., 119: 3510–3673 doi: 10.1021/acs.chemrev.8b00299.

[13] Watts, C., Crathorne, B., Fielding, M., Steel, C. (1983). "Identification of Non-Volatile Organics in Water Using Field Desorption Mass Spectrometry and High Performance Liquid Chromatography."In *Analysis of Organic Micropollutants in Water,* Edited by Angeletti, G. and Bjørseth, 120-131, A. Springer: Dordrecht.

[14] Richardson, M. L., &Bowron, J. M. (1985). "The Fate of Pharmaceutical Chemicals in the Aquatic Environment." *J. Pharm. Pharmacol.* 37:1–12. https://doi.org/10.1111/j.2042-7158.1985.tb04922.x.

[15] Rogers, I., Birtwell, I., & Kruzynski, G. (1986). "Organic Extractables in Municipal Wastewater Vancouver, British Columbia." *Water Quality Research J. Canada,* 21: 187–204. https://doi.org/10.2166/wqrj.1986.014.

[16] Hignite, C., & Azarnoff, D. L. (1977). "Drugs and Drug Metabolites as Environmental Contaminants: Chlorophenoxyisobutyrate and Salicylic Acid in Sewage Water Effluent." *Life Sci.,* 20: 337–341. doi: 10.1016/0024-3205(77)90329-0.

[17] Eckel, W. P., Ross, B., & Isensee, R. K. (1993). "Pentobarbital Found in Ground Water. Groundwater." *Groundwater,* 31: 801–804. https://doi.org/10.1111/j.1745-6584.1993.tb00853.x.

[18] Holm, J. V., Ruegge, K., Bjerg, P. L., & Christensen, T. H. (1995). "Occurrence and Distribution of Pharmaceutical Organic Compounds in the Groundwater Down gradient of a Landfill (Grindsted, Denmark)." *Environ. Sci. Technol.,* 29: 1415–1420. doi: 10.1021/es00005a039.

[19] Hirsch, R., Ternes, T., Haberer, K., & Kratz, K.L. (1999). "Occurrence of Antibiotics in the Aquatic Environment." *Sci. Total Environ.,* 225:109–118. doi: 10.1016/S0048-9697(98)00337-4.

[20] Cohen, J. (2018). "'Frightening' typhoid fever outbreak spreads in Pakistan." *Science.,* 361: 214. doi: 10.1126/science.361.6399.214.

[21] Maillard, J.Y., Bloomfield, S.F., Courvalin, P., Essack, S.Y., Gandra, S., Gerba, C.P., Rubino, J.R., & Scott, E.A. (2020). "Reducing antibiotic prescribing and addressing the global problem of antibiotic resistance by targeted hygiene in the home and everyday life settings: a position paper." *Am. J. Infect. Control.,* 48: 1090-1099. https://doi.org/10.1016/j.ajic.2020.04.011.

[22] Usman, M., Farooq,M., & Hanna, K. (2020). "Environmental side effects of the injudicious use of antimicrobials in the era of COVID-19." *Sci. Total Environ.,* 745: 141053.https://doi.org/10.1016/j.scitotenv.2020.141053.

[23] Geissen, V., Klumpp, M. H., Umlauf, E., Nadal, G., Van der Ploeg, M., Van de Zee, M., & Ritsema, S. E. (2015). "Emerging pollutants in the environment: a challenge for water resource management." 3: 57-65. https://doi.org/10.1016/j.iswcr.2015.03.002.

[24] Lee, C. M., Palaniandy, P., & Dahlan, I. (2017). "Pharmaceutical residues in aquatic environment and water remediation by TiO_2 heterogeneous photocatalysis: a review." *Environ. Earth Sci.,* 76, 17: 611. doi: 10.1007/s12665-017-6924-y.

[25] Samson, O.A., Ekemena, O. O., Olushola, S. A., Adam, A. S. Titus, A. M. Msagati., & Augustine, E. O. (2019). "Photocatalysis and biodegradation of pharmaceuticals, in wastewater: effect of abiotic and biotic factors." *Clean Technologies and Environmental Policy*: 1-23. https://doi.org/10.1007/s10098-019-01747-4.

[26] Alejandro, B., Majdi, B., Mokhtar, H., Elvira, G., Hatem R. A., Ramzi, D., & Albert, S. (2022). "Visible-light driven sono photocatalytic removal of tetracyclineusing Ca-doped ZnO nanoparticles." *Chem. Eng. J.,* 42: 132006. https://doi.org/10.1016/j.cej.2021.132006.

[27] Emanuela, C. G., Erica, C., Pengyun, L. Zhilin, W., & Giancarlo, C. (2021). "Degradation of Antibiotics in Wastewater: New Advances in Cavitational Treatments." *Molecules,* 2: 617. https://doi.org/10.3390/molecules26030617.

[28] ElShafei, G.M.S., Al-Sabagh, A.M., Yehia, F.Z., Philip, C.A., Moussa, N.A., Eshaq, G. & ElMetwally, A.E. (2021). "Metal oxychlorides as robust heterogeneous Fenton catalysts for the sonophotocatalytic degradation of 2-nitrophenol." *Appl. Catal. B Environ.,* 224: 681–691. https://www.sciencedirect.com/science/article/abs/pii/S0926337317310706.

[29] Moradi, M., Elahinia, A., Vasseghian, Y., Dragoi, E.-N., Omidi, F., & Khaneghah, A.M. (2020). "A review on pollutant removal by Sono-photo -Fenton processes." *J. Environ. Chem. Eng.* 8: 104330. https://doi.org/10.1016/j.jece.2020.104330.

[30] Andersson, D. I., & Hughes, D. (2014). "Microbiological effects of sublethal levels of antibiotics, AOP." *Nature Reviews Microbiology,* 12: 465-478. doi: 10.1038/nrmicro3270 .

[31] Kinsella, B., O'Mahony, J., Malone, E., Moloney, M., Cantwell, H., & Furey, A. 2009. "Current trends in sample preparation for growth promoter and veterinary drug residue analysis" *J. Chromatogram A.,* 1216, 46: 7977–8015. doi: 10.1016/j.chroma.2009.09.005.

[32] Van, B. T. P, Brower, C., Gilbert, M., Grenfell, B. T., Levin S. A., & Robinson, T. P. (2015). "Global trends in antimicrobial use in food animals" *Proc. Natl. Acad. Sci. U S A.*, 112:5649– 5654. doi: 10.1073/pnas.1503141112.

[33] Pruden, A., Joakim, L. D. G., Amézquita, A., Collignon, P., Brandt, K. K. & Graham, D. W., (2013). "Management options for reducing the release of antibiotics and antibiotic resistance genes to the environment." *Environ Health Perspect.,* 121: 878–885. doi: 10.1289/ehp.1206446.

[34] Svetlana, I. P., Anca, E. G., Bela K., Maria G. S., & Felicia, L. (2020). "Antibiotics in the environment: causes and consequences." *Medicine and Pharmacy Reports.*, 93: 231 – 240. doi: 10.15386/mpr-1742.

[35] Maria, P.,Miriam, S., Manuel, R., Maria-Inti, M., https://pubmed.ncbi.nlm.nih.gov/?term=Romero+CS&cauthor_id=34679203. Carolina, S. R., Agata, M., Peter, K. Patrick, M., Nicole, S., & Stephanie, W. (2021). "Antibiotics for the treatment of COVID-19." *Cochrane Database Syst. Rev.,* 22: 10: CD015025. doi: 10.1002/14651858.CD015025.

[36] Miranda, C., Silva, V., Capita, R., Alonso-Calleja, C., Igrejas, G., & Poeta, P. (2020). "Implications of antibiotics use during the COVID-19 pandemic: present and future." *J. Antimicrob. Chemothe,* 1-4. doi: 10.1093/jac/dkaa350 .

[37] Cong, W., Narayan Poudel, A. K., Alhusein, N., Wang, H., Yao, G., & Lambert, H. (2020) "Antimicrobial Use in COVID-19 Patients in the First Phase of the SARS-CoV-2 Pandemic: A Scoping Review." *Antibiotics.*, 10: 745. https://doi.org/10.3390/antibiotics10060745.

[38] Ana, B., Ana,S.,Josipa K. V., Ivan S., Marijan, A., Sanja, B., Tamara, J., Juan, J. G.alez P., Milena, M., & Nikolina U. K. (2017). "Negative environmental impacts of antibiotic-contaminated effluents from pharmaceutical industries." *Water Research.,* 126: 79-87, http://dx.doi.org/10.1016/j.watres.2017.09.019.

[39] Klaus,K. (2009). "Antibiotics in the aquatic environment – A review – Part II." *Chemosphere,* 75: 435–441. doi: 10.1016/j.chemosphere.2008.12.006.

[40] Zhang,T., Zhang, L., & Ma, C. (2020). "Degradation of Ceftriaxone Sodium in Pharmaceutical Wastewater by Photocatalytic Oxidation." *Biotechnology and Medicine,* https://webofproceedings.org/proceedings_series/ALSMB/WCCBM%202020/WCCBM20046.pdf.

[41] Joakim-Larsson, D. G. (2014). "Antibiotics in the environment." *Upsala Journal of Medical Sciences.*, 119: 108–112, doi: 10.3109/03009734.2014.896438.

[42] Underwood, J.C., Harvey, R.W., Metge, D.W., Repert, D.A., Baumgartner, L.K., Smith, R.L., Roane, T.M., & Barber, L.B. (2011). "Effects of the antimicrobial sulfamethoxazole on groundwater bacterial enrichment." *Environmental Science and Technology,* 45: 3096-3101. doi: 10.1021/es103605e.

[43] Antibiotics and the environment, The Minnesota Pollution Control Agency, https://www.pca.state.mn.us/featured/antibiotics-and-environment.

[44] Greluk, M., Gac, W., Rotko, M., Sowik, G., & Turczyniak-Surdackab, S. (2020). "$Co/CeO_2$and Ni/CeO_2catalysts for ethanol steam reforming: Effect of the cobalt/nickel dispersion on catalysts properties." *J. Catal.,* 393: 159–178. https://doi.org/10.1016/j.jcat.2020.11.009.

[45] Rui, N., Zhang, X. S., Zhang, F., Liu, Z. Y., Cao, X. X., & Xie, Z. H. (2021). "Highly active Ni/CeO_2catalyst for CO_2methanation: Preparation and characterization." *Appl. Catal.,* 282: 119581. https://doi.org/10.1016/j.apcatb.2020.119581.

[46] Palacio, R., Lopez, D., & Hernandez, D. (2019). "Bimetallic Au Cu nano particles supported on CeO_2 as selective catalysts for glycerol conversion to lactic acid in aqueous basic medium." *J. Nanopart. Res.,* 21, 7: 1–13. doi: 10.1007/s11051-019-4594-2.

[47] Feng, N. J., Zhu, Z. J., Zhao, P., Wang, L., Wan, H., & Guan, G. F. (2020). "Facile fabrication of trepang-like CeO_2@MnO_2 nanocomposite with high catalytic activity for soot removal." *Appl. Surf. Sci.,* 515: 146013. https://doi.org/10.1016/j.apsusc.2020.146013.

[48] Zeng, Y. Q., Haw, K., Wang, Z. G., Wang, Y. N., Zhang, S. L., & Hongmanorom, P. (2020). "Double redox process to synthesize CuO–CeO_2 catalysts with strong Cu–Ce interaction for efficient toluene oxidation." *J. Hazard. Mater.,* 404: 124088. https://doi.org/10.1016/j.jhazmat.2020.124088.

[49] Yang, X. Y., Lei, X. X., & Tang, S. K. (2020). "Ionic liquid-assisted synthesis of hierarchically olive-like $Al_2O_3$and CeO_2/Al_2O_3 with enhanced catalytic activity for Orange II degradation." *Ceram. Int.*, 47, 4: 5446–5455. https://doi.org/10.1016/j.ceramint.2020.10.126.

[50] Moogi, S., Lee, I., &Kee, H. (2020). "Catalytic steam reforming of glycerol over Ni–La_2O_3–CeO_2/SBA-15 catalyst for stable hydrogen-rich gas production." *Int. J. Hydrogen Energy*, 45, 53: 28462–28475. https://doi.org/10.1016/j.ijhydene.2020.07.249.

[51] Zhao, W. J., Zhang, Z. C., Zhang, J., Wu, H. G., Xi, L. M., & Ruan, C. H. (2016). "Synthesis of Ag/TiO_2/graphene and its photocatalytic properties under visible light." *Mater. Lett.*,171: 182–186. https://doi.org/10.1016/j.matlet.2016.02.063.

[52] Qin, Y., Qu, Z. P., Dong, C., & Huang, N. (2017). "Effect of pretreatment conditions on catalytic activity of Ag/SBA -15 catalyst for toluene oxidation." *Chin. J. Catal.,* 38, 9: 1603–1612. DOI:10.1016/S1872-2067(17)62842-0.

[53] Nasrollahzadeh, M., Issaabadi, Z., & Sajadi, S. M. (2019). "Green synthesis of the Ag/Al_2O_3 nanoparticles using *Bryonia alba* leaf extract and their catalytic application for the degradation of organic pollutants." *J. Mater. Sci.: Mater. Electron*., 30: 3847–3859. https://doi.org/10.1007/s10854-019-00668-8.

[54] More, P. M. (2017). "Effect of active component addition and support modification on catalytic activity of Ag/$Al_2O_3$for the selective catalytic reduction of NOx by hydrocarbon – A review." *J. Environ. Manage.,*188: 43–48. https://doi.org/10.1016/j.jenvman.2016.11.077.

[55] Gao B., Chakraborty A.K., Yang J. M. & Lee W. I. (2010). "Visible-light photocatalytic activity of BiOCl/Bi_3O_4Cl nanacomposites." *Bull. Korean Chem. Soc.* 31, 7: 1941-1944. https://doi.org/10.5012/bkcs.2010.31.7.1941.

[56] Sanjay, K. S. (2018). "Green Chemistry and Sustainability Research Group, Department of Chemistry, JECRC University, Jaipur, Rajasthan, India." doi: https://doi.org/10.1016/C2018-0-01479-6.

[57] Ruppert, G., Bauer, R., & Heisler, G. (1993). "The photo-Fenton reaction—an effective photochemical wastewater treatment process." *J. Photochem. Photobiol. A: Chemistry,* 73, 1: 75-78. https://doi.org/10.1016/1010-6030(93)80035-8.

[58] Gurol, M.D., & Vatistas, R. (1987). "Oxidation of phenolic compounds by ozone and ozone or UV radiation: a comparative study." *Water Research*, 21, 8: 895-900. https://doi.org/10.1016/S0043-1354(87)80006-4.

[59] Sharma, A., Ahmad, J., & Flora, S.J.S. (2018). "Application of advanced oxidation processes and toxicity assessment of transformation products." *Environmental Research,* 1-36. doi: https://doi.org/10.1016/j.envres.2018.07.010.

[60] Takahashi, N. (1990). "Ozonation of several organic compounds having low molecular weight under ultraviolet irradiation" *J. Int. Ozone Association*, 12: 1-27. https://doi.org/10.1080/01919519008552452.

[61] Nguyen, Q. T., Amr, S., Li-Chyong, C., Kuei-Hsien, C., Cao Minh, T., & Pham Van, V. (2021). "High-efficient photocatalytic degradation of commercial drugs for pharmaceutical wastewater treatment prospects: A case study of Ag/ g-C_3N_4/ZnO nanocomposite materials." *Chemosphere,* 282: 130971. https://doi.org/10.1016/j.chemosphere.2021.130971.

[62] Marta, I. L., & Natalia, Q. (2010). "Photochemical Advanced Oxidation Processes for Water and Wastewater Treatment." *Recent Patents on Engineering*., 4: 217-241. doi: 10.2174/187221210794578574.

[63] Manoj, P., Rayaroth, E. E., & Yoon-Seok, C. (2020). "Degradation studies of halogenated flame retardants." *Comprehensive Analytical Chemistry*, 88: 303-339. https://doi.org/10.1016/bs.coac.2019.11.003

[64] Han, Q., Dong, W., Wang, H., Ma, H., Liu, P., Gu, Y., Fan, H., & Song, X. (2019). "Degradation of tetrabromobisphenol by ozonation: performance, products, mechanism and toxicity." *Chemosphere,* 235: 701–712. doi: 10.1016/j.chemosphere.2019.06.204.

[65] Huang, J., Wang, X., Pan, Z., Li, X., Ling, Y., & Li, L. (2016). "Efficient degradation of perfluorooctanoic acid (PFOA) by photocatalytic ozonation." *Chem. Eng. J.,*296: 329–334. https://doi.org/10.1016/j.cej.2016.03.116.

[66] Anastas, P. T., & Warner, J. C. (1998). *Green Chemistry: Theory and Practice.* Oxford University Press: New York. https://www.worldcat.org/title/green-chemistry-theory-and-practice/oclc/39523207.

[67] Vinodgopal, K., Bedja I., Hotchandani, S., & Kamat, P. V. (1994). "A Photocatalytic Approach for the Reductive Decolorization of Textile Azo Dyes in Colloidal Semiconductor Suspensions." *Langmuir,* 10: 1767-1771. https://doi.org/10.1021/la00018a024.

[68] Zou, R., Angelidaki, I., Yang, X., Tang, K., Rasmus, A. H., & Zhang, Y. (2020). "Degradation of pharmaceuticals from wastewater in a 20-L continuous flow bio-electro-Fenton (BEF) system." *Science of the Total Environment,* 727: 138684, https://doi.org/10.1016/j.scitotenv.2020.138684.

Chapter 10

A Critical Review on Treatment of Pharmaceutical Industrial Effluents in Waterbodies

Susmita Bajpai[1,*]
Department of Chemistry, Brahmanand College, Kanpur, India

Abstract

Emerging pollutants in the environment, such as medicines comprising of harmful chemicals are causing increasing worry throughout the world. The increased use of medication across the world along with pharmaceutical resistance to typical wastewater treatment technologies has resulted in a serious environmental hazard. Pharmaceutical compounds are often made in batch processes resulting in a diverse range of products in created wastewater. The large amount of created wastewater is utilized in various activities like solid cake washing, extraction, or equipment cleaning. Drug availability, access, and usage have all increased in recent years, causing significant increase in drug concentrations in water bodies.

Given the limitations of water supply as a part of water management, it is vital to comprehend and establish approaches for the treatment of pharmaceutical wastewater. The different sources of wastewater in the pharmaceutical sector are outlined in this chapter and the best current solutions to remove them are critically assessed. Pharmaceutical wastes which are considered emerging contaminants pose a challenge in the field of environmental remediation. As a result, alternative add-on systems for traditional wastewater treatment plants are being developed all the time

* Corresponding Author's Email: susmitabajpai999@gmail.com.

In: Pharmaceuticals: Boon or Bane
Editor: Neha Agarwal
ISBN: 979-8-88697-487-4

to mitigate their impact and reduce their effects on the environment and human health. We present the current state and effect of pharmaceutical chemicals as emerging pollutants in this chapter. Examining the development of bioremediation systems particularly mycoremediation for the removal of these pharmaceutical substances with a special focus on fungal technologies and also focusing on their presence in the water bodies. The recommendations made in this chapter will be beneficial in the treatment of pharmaceutical industrial effluents.

Keywords: pharmaceuticals, mycoremediation, water management, effluents

1. Introduction

Water is essential for both humans and wildlife, and the world's functioning is dependent on the supply of clean water. Ground and surface water quality monitoring is essential since they are the primary sources of water for home and industrial usage. Water bodies have recently been damaged by emerging contaminants (ECs), which have the ability to enter the ecosystem and have a negative influence on human health and the environment. The water quality indicators used to evaluate watercourses receiving industrial effluents were deemed adequate. The adaptability of these indices in relation to the parameters used in the calculations facilitates water quality evaluation in developing countries. In these countries, high costs and a lack of necessary infrastructure for other parameters are currently deterrents to appropriate site-specific water quality evaluation.

Pharmaceutical compounds are released into drinking water from two sources: pharmaceutical industry manufacturing processes and general usage of pharmaceutical compounds, which results in their presence in urban and agricultural wastewater. A vast range of chemicals can be found in wastewater created during various procedures in the creation of medicines and medications. Furthermore, industry mandates the reuse of water following the removal of impurities, whether pharmaceuticals or otherwise, from effluents of various sectors of active pharmaceutical ingredients (API), bulk drugs, and related pharmaceutics. All of which use large amounts of water, are evaluated and strategies are made for recovering valuable compounds to a large extent are proposed. This is followed by a discussion of the treatment of very dilute but harmful wastewater. Pharmaceuticals cannot be entirely removed from wastewater by any singular approach. The optimum solution appears to be a hybrid wastewater treatment system that combines traditional treatment methods with membrane reactors and sophisticated post treatment procedures. While medications offer unquestionable advantages, governments are still grappling with how to handle their potential consequences on drinking water quality, ecology and human health. Pharmaceutical pollution puts traditional water quality management to the test, necessitating novel wastewater treatment technology as well as behavioural adjustments in business, agriculture, health care, and society at large. Pharmaceutical contaminants (PCs) are biologically active substances intended to prevent, cure, or treat illnesses and are one of the most concerning kinds of ECs that come from the pharmaceutical industry. The pharmaceutical industry has contributed significantly to strong economic growth over the last two decades. It is India's fourth biggest manufacturer and accounts for around 20% of the country's GDP in terms of volume and jobs. From pharmaceutical production, there is a large quantity of waste dumped into water bodies, which causes significant water pollution. The pharmaceutical sector produces waste that contains substances that are toxic to humans and aquatic life. The produced wastewater reduces soil fertility, increases hardness of water, salinity and enhances resistance to germs in water [1].

Current water quality policy approaches are usually reactive, interventions are implemented only when routine monitoring is in place and hazards can be demonstrated. Many national approaches to the pharmaceutical crisis have centred on surveillance and end-of-pipe safeguards (e.g., upgrading wastewater treatment plants and implementing public take-back programmes for unused and expired medicines). Upstream, source-directed and use-oriented initiatives, such as antibiotic limits in agriculture, green pharmacy, and public environmental health campaigns, are emerging. Anthropogenic pollution of river flows generated by activities such as industry, agriculture, and human waste is a major problem. It may have large-scale negative effects on surface water quality when combined with an increasing population and climate change. The use of water quality indices (WQI) streamlines the presenting of results from any analysis connected to a water body by summarizing a series of measured factors into a single number or concept. The index used is determined by the pollution sources existing at the area and the anticipated use of the water, as well as the choices for doing the necessary study.

To meet the demands of ever-stricter regulatory discharge restrictions, the pharmaceutical sector requires constant high-quality water for manufacturing and wastewater treatment. To overcome these issues different industries must question conventional thinking and tactics while also exploring new technology and solutions in order to remain competitive. Thus, attempts have been made in the current chapter to –

1. Understand the nature of pharmaceutical waste generated at the industry site
2. Classify the waste generated by various industrial processes
3. Evaluate the efficacy of advanced processes and hybrid technologies for removing pharmaceutics from aqueous systems.

This chapter investigates the properties of pharmaceutical industry effluents, eventually leading to the creation of a trickling bed reactor for waste treatment utilising *Bacillus subtilis*. Industrialisation has played a significant role in the advancement of economic growth during the last few decades. This has resulted in a considerable volume of waste material being discharged into water bodies, which has increased the prevalence of many illnesses. As a result, many developing countries have manufactured a diverse range of medications. Along with the processing, a large amount of pharma waste is produced and disposed off in the environment. According to the chapter the presence of pharmaceutical chemicals in water bodies and over 30 distinct compounds have been identified in amounts larger than the detection limit in numerous Asian nations.

The release of pharmaceutical substances into the aquatic environment, as well as the pollution risks associated with it has become a serious concern for the pharmaceutical industry. The long-term exposure of medications in water bodies causes chronic and acute poisoning. Pharmaceutical wastewater can modify behaviour and potentially affect population and community structures. The ecological fitness of an organism is generally indicated by its behavioural reactions. Changes in behaviour together with changes in physiology, may affect population stability. Several investigations have found that aquatic species exposed to pharmaceutical effluents experience reproductive problems [2, 3].

The sequence of operations in the pharmaceutical sector results in the formation of harmful wastewater and its discharge into water bodies. The wastewater contains solid as well as biodegradable and non-biodegradable organic pollutants. The physical and chemical characteristics of waste should always be within the pollution control board's regulatory limits. Another important factor is the amount of oxygen in the effluent. It is derived from the COD (chemical oxygen demand) and BOD (biochemical oxygen demand). The remaining nutritional status is determined by the amount of nitrogen and phosphorus in waste water. Other water quality characteristics are classified as conductivity, hardness, and turbidity. The chemical and physical treatment of pharmaceutical wastewater has resulted in more complicated contamination and the inclusion of complex by-products making treatment more difficult and polluting aquatic bodies. Not only that but they also require chemicals and treatment processes that are exceedingly expensive, and, at that time, difficult to handle at the industrial level. Many treatment facilities use biological processes to remove N, P, and organic waste. This is mostly owing to the lower cost and simpler procedure. Multiple factors and characteristics impact biological treatment, many of which cannot be modified since they are related to the entering wastewater such as organic matter, nutrients, trace elements, toxins, and so on.

Trickle bed reactors are three-phase reactors that add a trickling effect to packed bed reactors. The gas and liquid travel over the solid substrate either concurrently or counter currently. The trickling effect has been added to allow aerobic bacterial activity to remove organic materials from wastewater streams. In general, they may be run with or without recycling. Recycling is recommended in large-scale operations because it provides better flow distribution with larger loads. However, it is more crucial than in submerged packed bed operation. This reactor also has minimal pressure drop is easy to operate and build, has fixed components and has low liquid and air flow rates.

The pharmaceutical sector produces wastewater that contains substances that are toxic to humans and aquatic life [4]. As a result, they cannot be released directly into municipal bodies. During wastewater treatment, however, sewage treatment services may not adequately eliminate all elements. The wastewater produced enhances resistance to germs in water. As a result, there is a great deal of concern about pharmaceutical industry effluents and its treatment. As a result, this chapter primarily focuses on examining the features of pharmaceutical industrial effluents and eventually constructing a reactor to neutralise the effect while also bringing the characteristics within acceptable ranges.

2. Types of Wastewater

In general, wastewater is divided into two categories: sewage wastewater and non-sewage wastewater. Domestic wastewater is included in sewage wastewater. Sewage wastewater is wastewater that contains bodily wastes (urine and faeces) and is generated by places such as households, schools, hospitals, hotels, restaurants, public toilets, and so on. Non-sewage wastewater refers to all other forms of wastewater created by commercial activities, such as those produced by factories and industrial facilities. Stormwater and rainwater created during rainfall or flood events are also included in non-sewage wastewater. Figure 1 represents the division of different types of waste waters generated by industrial and non-industrial sectors.

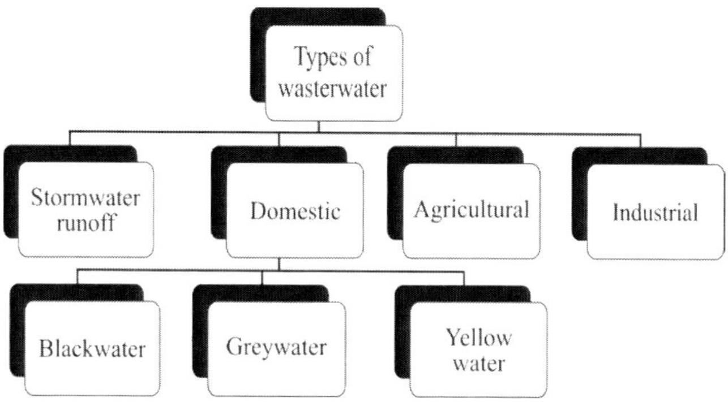

Figure 1. Types of wastewater generated by various industrial and non-industrial sectors.

Human activities are heavily reliant on water, making wastewater management and treatment critical. As a result, for effective management and targeted treatment, wastewater has been further classified into well-defined kinds and sub-types based on its source. Figure 1 depicts the four primary forms of wastewater: stormwater runoff, household, agricultural, and industrial. All of these categories and their sub-types are covered briefly below.

2.1. Stormwater Runoff Wastewater

Heavy rainfall, storm, or flood water that does not seep into the earth and runs over the street or open surfaces is referred to as stormwater runoff wastewater. Many harmful pollutants, such as plastics, pesticides, herbicides, oils, chemicals, heavy metals, and even numerous diseases, are carried off into stormwater runoff from roadways, industrial sites, construction sites, and other locations, making it one of the primary contributors of water pollution. Stormwater runoff often travels directly or through channelled drains, eventually discharging untreated into neighbouring natural waterways such as ponds, rivers, streams, and lakes. This filthy water not only harms aquatic life, but it also endangers the overall ecosystem because all living forms are directly or indirectly linked to natural rivers.

2.2. Domestic Wastewater

Domestic wastewater is wastewater generated by household activities. The two major waste streams that contribute to this wastewater are toilet waste (liquid produced from sanitary, laundry, and bathing facilities) and wastewater created by other home activities such as cooking. Domestic wastewater is classified into three categories based on the source: black, grey, and yellow wastewater.

2.2.1. Blackwater

The most polluted type of household wastewater is that which is discharged from toilets, dishwashers, and sinks. Urine, faeces, toilet paper, soaps, leftover food items, different chemicals, and a variety of cleaning products are the pollutants found in blackwater. It is very filthy wastewater with a significant potential for disease transmission.

2.2.2. Greywater

Domestic wastewater that is discharged from bathtubs, washing machines, and bathroom sinks is less polluted. To put it simply, greywater or sullage in blackwater that does not contain faeces, urine, or particles of food waste, i.e., domestic or household wastewater that has not come into touch with toilet water. Though it is not pathogenic, because it contains detergents, soaps, cleaning products, and other chemicals, it should be well cleaned before reuse.

2.2.3. Yellow Water

This is only urine with no extra impurities from blackwater or greywater. Yellow water has no faeces, toilet paper, chemicals, or even food particles and is therefore pure pee water. Domestic wastewater classification simplifies the treatment planning and execution since unique treatments may be given to the kind of water depending on its characteristics.

2.3. Agricultural Wastewater

Agriculture runoff is a significant cause of water contamination in many watersheds. When surplus water rushes off the fields during surface irrigation, it is commonly referred to as irrigation tailwater. The surplus water flowing through the fields is the principal source of silt and nutrient overflow into surrounding water sources. Furthermore, agrochemicals such as fertilisers, pesticides, herbicides, crop residues, animal wastes, pig, poultry, and fish farm effluents, and dairy farming waste are agricultural wastewater pollutants. To significantly reduce surface runoff, many farm management approaches are employed for agricultural wastewater treatment.

2.4. Industrial Wastewater

Industrial wastewater is defined as water containing dissolved and suspended pollutants discharged from various industrial operations, such as water produced during manufacturing, cleaning, and other commercial activities. The pollutants found in industrial wastewater vary depending on the plant and sector. Mining, steel/iron production facilities, industrial laundries, power plants, oil and gas fracking operations, metal finishers, and the food and beverage business are all examples of industries that generate wastewater. Chemicals, heavy metals, oils, pesticides, silt, medicines, and other industrial by-products are among the pollutants often detected in industrial water outflows. In general, treating industrial wastewater is tough because unique analysis of set-ups and particular treatment facilities is necessary on an industry-by-industry basis. As a result, to address this, on-site filter presses are installed to treat the effluent wastewater.

3. Wastewater and Its Risk to Human Health

The increasing and widespread use of wastewater in agriculture and other applications, with or without inadequate treatments, poses serious public health risks that must be addressed. However, there are well-developed and advanced wastewater treatment technologies, but they do not appear to be sustainable for low-income countries due to costly investment and technological hurdles [5, 6]. As a result of the course that wastewater takes through households, factories, and other sites of use, it may include various pathogenic microbes, chemical and pharmaceutical residues, and pose health hazards if not properly mitigated. Salts, metals, metalloids, residual medicines, organic chemicals, endocrine disrupting substances, and active residues of personal care items are examples of chemical pollutants [7, 8]. Several aspects, such as wastewater treatment extent, pollutant characteristics, human exposure, and local risk factors, determine the kind and intensity of health problems encountered. Infectious outbreaks continue to be a major concern in low-income nations, while chemical and pharmaceutical pollution is a serious health danger in both developing and high-income or developed countries [9, 10].

Some underdeveloped countries confront nearly all sorts of health difficulties in this context. The majority of pharmaceutical manufacturing factories and academic/research institutes are located inside the city and are dedicated to the synthesis of novel nanomaterials, chemicals, and medications. Untreated or partially treated wastewater contains various chemicals and heavy metals; on the other hand it may find its way into certain local drinking water wells. The situation becomes more stressful when we learn that most cities lack a complete wastewater treatment plan, and untreated wastewater is utilised for irrigation to grow vegetables for direct human consumption [11, 12, 13, 14, 15, 16, 17].

Wastewater can be used for irrigation, groundwater supplies, restoration, industry, the environment, potable and non-potable municipal usage, and so on. Although wastewater consumption in European nations is primarily focused on agriculture, industry, municipal, and mixed applications, low or untreated wastewater is widely utilised for agricultural irrigation in low and moderate income countries [5, 17]. The public health risk might be handled by programmes that include the removal of pathogenic microorganisms in wastewater, the removal of chemical pollutants, and the reduction of human exposure to wastewater. The water quality approach is a major risk management approach advocated by the World Health Organization (WHO) and the Environmental Protection Agency (EPA), which states that wastewater, should be treated to the extent that it meets certain water quality criteria in order to avoid any related risk, particularly health concern [7]. This approach encourages nations to examine their own cultural, social, and environmental factors when developing and implementing long-term, cost-effective risk management initiatives [18]. However, it has been widely agreed that the amount of wastewater treatment should correspond to the aim of reuse. As a result cost-effectiveness studies are critical for evaluating various risk management choice for evidence-based decision-making, intervention selection, and resource allocation objectives, particularly in low-income and developing countries [5]. The WHO pathogen reduction guidelines are based on a tolerable increased disease burden of one disability adjusted life year loss per person per year, and they include recommendations for pathogen reduction strategies such as:

- Health protection for wastewater-irrigated field workers against the danger of viral, bacterial, and protozoan infections, with wastewater treatment achieving a pathogen reduction of 3-4 log units.
- Consumers of wastewater-irrigated food crops should be protected against the danger of viral, bacterial, and protozoan infections by achieving a 6-7 log unit pathogen reduction by wastewater treatment and subsequently post-treatment human health-protection control activities.
- Health protection for wastewater-irrigated field workers and consumers of wastewater-irrigated food crops against the danger of helminthic infections by reducing the human intestinal nematode egg count to less than one per person per litre [5, 7].

4. Treatment of Industrial Wastewater

Various types of technologies and strategies are being developed and employed for the removal of contaminants from wastewater released by several industries. Some of the strategies developed and used by some major wastewater-producing industries for treatment of effluents are mentioned below.

4.1. Containing Heavy Metals

Heavy metals are one of the most dangerous contaminants emitted by chemical-intensive industries. To remove heavy metals from inorganic effluents, conventional techniques like ion-exchange methods using synthetic ion-exchange matrices for cation and anion exchange, chemical precipitation using precipitants like lime and limestone under basic pH conditions, and electrochemical deposition methods are being used. Many drawbacks are associated with the complete removal of heavy metals using these procedures and they also demand a lot of energy [19]. Adsorption, membrane filtration, electro-dialysis, and photocatalysis are a few of the less expensive and efficient methods created for improving the quality of treated water. The removal of heavy metals can be done efficiently using low-cost adsorbents like natural materials e.g., zeolites and clinoptilolite, industrial by-products like iron slags, fly ash, hydrous titanium oxide, and waste iron. Biosorption using biological and agricultural wastes like inactive microbial biomass, orange peel, pecan shells, hazelnut shells, maize husk or cob, etc., and modified biopolymers like chito membrane filtration methods, such as ultrafiltration, that use permeable membranes with pore sizes of between 5 and 20 nm can be used to remove heavy metals from inorganic solutions. Other methods for removing heavy metals include polymer-supported ultrafiltration, nanofiltration, and reverse osmosis, which can remove 98 percent of copper and 99 percent of cadmium [20]. The ionised solution is transported through an ion-exchange membrane during the electrodialysis process, and membrane separation occurs at an electric potential. Heavy metal ions like Ni, Co, and Cd can be effectively removed with this approach. The photocatalysis approach, on the other hand, makes use of titanium dioxide semi-conductors that may reduce or oxidise species with the right redox potential, such as Cu^{2+}, Cr^{3+}, and Cr^{4+} heavy metal ions [21].

4.2. Containing Phenolic Compounds

One of the most common refractory chemical contaminants found in wastewater discharged by businesses is phenol and phenolic compounds [22]. Chemical, physical, electrochemical, and anaerobic biological techniques can be used to remediate phenolic wastes. According to reports, the electrochemical method is the most efficient way to eliminate phenolic waste [23]. For the direct or indirect oxidation procedures used to destroy the contaminants, electrons are the primary reagent. In the electrochemical treatment of tannery wastewater, landfill leachate, resorcinol, and cresols [24], Ti/Pt & Ti/Pt/Ir, graphite, $Ti/SnO_2–PdO_2–RuO_2$, & $TiO_2–RuO_2–IrO_2$ anodes have been described. Activated carbon is used to adequately treat effluents before they are released into the environment because electrochemical treatment raises the concentration of adsorbable organically bound halogens (AOX) in thc cffluents [25].

4.3. Released from the Paper and Pulp Industry

This industry is one of the most water-exhaustive and highly polluting industrial sectors, which along with liquid waste also generates a huge amount of solid and gaseous waste. A number of treatment processes are used to manage this waste. The black liquor produced is treated by a membrane filtration system by vibration separation enhanced processing (VSEP) and biological treatment, AOX is reduced by oxygen bleaching, heavy metals are treated by biological and sedimentation treatment methods. Anaerobic digestion (AD) followed by pyrolysis and incineration is mainly used for primary and secondary bio sludge treatment. Anaerobic digestion with incineration helps in the production of beneficial by-products like biochar and biogas and also helps in removal of lignin. Advanced oxidation with Fenton's reagent, ozonation, gasification and biological treatment are also employed to treat wastewater. A composite coagulant, polymeric ferric aluminium sulphate chloride (PFASC) works with polyacrylamide (PAM) as a tertiary treatment of wastewater from paper mills that reduces chroma (intensity of colour released in paper mill wastewater), COD, and BOD to a great extent. The high content of BOD and COD in wastewater is treated by aerobic granulation, which helps in the removal of tannin and lignin [26]. The chroma produced by the paper and pulp industies can also be reduced by use of agroindustry-based residual biosorbents such as activated carbon and agricultural by-products. The latest approach is the use of microbial fuel cells for the treatment of wastewater generated by industries. It has benefits like electrical energy production and exclusion of the aeration process, which is conventionally used for the removal of many dissolved gases. The alternative eco-friendly use of enzyme bleaching, (xylanase and laccase) instead of chlorine bleaching reduce the production of organic chlorinated pollutants.

4.4. Released from the Textile Industry

Additionally, the textile sector uses a lot of freshwater. Since water is needed for many processing steps, a lot of wastewater is produced as a result. The wastewater produced by the textile industry also includes additional wastes, such as dyes (azo dyes) used for colorization. Adsorption of membrane-based separation techniques and ion-exchange procedures are examples of traditional physical-chemical treatment processes [27]. For the elimination of dyes, adsorbents such as silicon, carbon and kaolin polymers are utilised. Reactive dyes and other chemical compounds are treated in water using membrane separation techniques including nanofiltration and reverse osmosis, while anionic and cationic dyes are removed from wastewater using the ion-exchange approach. The Fenton's reagent, a potent oxidising agent with sufficient hydrogen peroxide added for decolourization is another common technique in addition to these. Ozonation is used for hazardous and non-biodegradable components. By using UV light photochemical processes destroy dyes. Basic, acidic, reactive and dispersion dyes can all be completely degraded when treated with cucurbituril, a polymer of formaldehyde and glycoluril. In addition to these traditional approaches, biological treatment using various microbes, bacteria, fungi, algae, and plants has proven to be highly efficient in the breakdown of these chemically stable pigments. Reductases are used during anaerobic bacteria treatment to break down the azo linkage of the dyes by *Pseudomonas putida, Staphylococcus hominis, & Citrobacter* species. Under aerobic conditions, species with high substrate specificity, such as *Geobacillus stearothermophilus, Micrococcus* sp. *and Staphylococcus arlettae,* develop oxygen-insensitive azo-reductases that employ NADH as a co-factor to cleave the azo linkages. A treatment system with a mixed culture of bacteria is also used for a more advanced biodegradation of these contaminants from wastewater. By creating non-specific enzymes like laccase, lignin, and manganese peroxides for the destruction of azo-dyes, fungi strains like *Phanerochaete chrysosporium* decolorize materials. By utilising azo-dyes as a source of nitrogen, algae aid in the breakdown of these dyes and help to keep water bodies free from becoming overly eutrophicated. By utilising methods like bioadsorption and biodegradation microalgae are employed to remediate the effluents emitted by the textile sector. Textile effluents are treated using phytoremediation, which uses methods like phytotransformation, phytostimulation, phytovolatilization, phytoaccumulation, rhizofiltration, and phytostabilization.

4.5. Treatment of Hypersaline Effluents

Hypersaline effluents, which are released by some industrial sectors, can be treated by physico-chemical methods. These methods involve thermal techniques that use multiple effect evaporators (MEE). This reduces the volume of effluent and leads to the separation of a solid salt. A coagulation–flocculation method is used as a pre-treatment for removal of the colloidal

COD fraction from hypersaline effluents. Some other effective techniques applied for desalination are ion-exchange methods using both anionic and cationic exchangers and membrane filtration techniques like reverse osmosis and electrodialysis.

5. Emerging Policy Instruments for Controlling Discharge of Pharmaceutical Instruments

As detection and understanding of pharmaceuticals in the environment improves, as does interest of the scientific community and the public media, so raises the need for policy intervention to mitigate human and environmental dangers. Various options for reducing pharmaceuticals in the environment have received minimal attention; several OECD nations and research institutes have concentrated on monitoring programmes, issue identification, risk analysis, and prioritisation of hazardous medicines. This chapter examines policy approaches to the problem that go beyond monitoring, reporting, and risk assessment. Ongoing national policy responses and country case studies are documented and classified as follows:

I. source-directed methods
II. use-oriented approaches
III. end-of-pipe pharmaceutical management strategies

5.1. Source Directed Methods

Source-directed policy methods encompass a variety of policy instruments that enforce, incentivize, or promote steps to prevent dangerous compounds from entering into the water bodies and having a negative impact on aquatic ecosystem. They are primarily aimed at pharmaceutical companies and manufacturing facilities but regulatory bodies, companies that buy bulk drugs, international investment firms, pharmacies, and healthcare organisations that buy and distribute finished pharmaceutical products have the ability to influence and create incentives to manage their release [28]. This chapter summarises a variety of source-directed regulatory, economic, and voluntary policy tools affecting a wide range of intersecting policy sectors, including health, water, agriculture, industry, and environmental protection.

The sections that follow, go into further depth on some of these policy tools and give case study examples of their implementation. Source-directed techniques also reduce the need for and expense of end-of-pipe wastewater treatment modifications and solid waste disposal while promoting drinking water safety. Green pharmacy, sound manufacturing practises, EQNs and water quality standards and water safety planning are a few examples. Another source-directed policy approach is substance prohibition. Diclofenac has been restricted for veterinary use in India, Nepal, and Pakistan due to the fall of vulture populations in the Indian subcontinent. To preserve the environment, the German Environment Agency suggests that veterinary medications containing active chemicals with PBT characteristics should be banned [29]. When a chemical prohibition is discussed in politics one must assess the cost of employing them for what may be beneficial to human and animal health advantages in comparison to the cost of harm if usage is prolonged. It is also necessary to investigate (analyse alternatives) whether a lower-risk alternative medicine is available (i.e., one that does not simply result in pollution-swapping).

5.2. Use Oriented Approach

Policy instruments that mandate, incentivize, or promote a reduction in the use of medicines and their release into the environment are examples of usage-oriented policy initiatives. They are intended to educate and influence physicians, veterinarians, pharmacists, patients, and farmers to change their attitudes and behaviours. For example, reducing inappropriate and excessive pharmaceutical consumption can be accomplished through: regulatory instruments, such as restrictions on over-the-counter sales of pharmaceuticals, that are harmful to the environment and mandatory codes of conduct for health practitioners, veterinarians, and farmers; ii) economic instruments, such as government subsidies in exchange for private sector environmental commitments; and iii) voluntary measures such as pharmaceuticals.

Use-oriented pharmaceutical policy tools have only recently been studied in general; they entail the alteration of long-established standards in clinical prescribing, human and animal health practises. However, because excretion and release through WWTPs is one of the major pathways of medicine into the environment, use-oriented techniques may be more effective and less expensive than end-of-pipe treatments [30]. Instead of relying solely on medicines, there are prospects for new economic models in the health care and agriculture industries focused on enhanced service and education. When considering a drug prohibition, one must assess the cost of employing them for what may be beneficial human and animal health advantages in comparison to the cost of harms if usage is sustained. It is also necessary to examine (assess alternatives) whether a less hazardous alternative medication is available (i.e., does not simply result in pollution-swapping).

5.3. End-of-Pipe Pharmaceutical Management Strategies

End-of-pipe methods aim to eliminate medications after they have been used or released into water. End-of-pipe policies include a variety of tools that mandate, incentivize or promote better wastewater treatment and solid waste disposal. One policy approach for reducing pharmaceuticals in water is to give incentives for the upgrade of WWTPs with new removal technology. Conventional WWTPs are not designed to remove pharmaceuticals and don't do so completely. Another alternative is to enhance solid waste collection and disposal in order to keep unwanted medications out of the water cycle. On their own, end-of-pipe solutions will not solve the problem of medicines in water. They are constrained by low removal efficiency, high capital and operating costs, and increasing energy consumption. Furthermore, end-of-pipe treatments do not address diffuse pharmaceutical contamination sources (e.g., from agriculture and aquaculture). However, because medications are still required for health and well-being, end-of-pipe interventions will remain a policy choice. End-of-pipe regulatory, economic, & voluntary policy tools are summarised in Table 1.

It affects a wide range of stakeholders, including customers, wastewater utilities, solid waste collectors, the pharmaceutical industry, and health and environmental regulators. The subsections that follow examine some of these policy instruments in further depth and offer case study examples of their implementation. Much of this section focuses on the removal of pharmaceutical residues using advanced wastewater treatment, its cost-effectiveness, and prospective financing sources to support WWTP improvement.

Table 1. End-of-pipe policy instruments to remove pharmaceuticals after their use and release into water bodies

Policy instrument type	Policy instrument	Description
Regulatory	Best available technique (BAT)	Definition of the best technology options for improved wastewater treatment
	Wastewater treatment standard	Definition of performance standards for wastewater treatment (e.g., treatment capacity or effluent load) without requiring a specific technology
	Pharmaceutical disposal requirements	Standard on correct waste disposal, e.g., mandatory consumer-level "take-back" programmes for unused pharmaceuticals
Economic	Effluent/ emission charges	Tax on discharging wastewater to water bodies, in order to incentivise emission reduction
	Wastewater tariffs or taxes for WWTP upgrades	Tariffs or taxes designed to signal the cost of wastewater treatment to remove pharmaceuticals to the public and consumers
	Subsidies for improved wastewater treatment	Financial support from governments to incentivise operators to invest in advanced wastewater treatment; or to promote research on improved wastewater treatment
	Extended producer responsibility (EPR) schemes	Instead of consumers being responsible for the cost of wastewater and waste management, producers, to some extent, become responsible for financing the end-of-life costs (wastewater treatment and solid waste disposal). In principle, companies can internalise these costs and are incentivised to produce pharmaceuticals more cost-efficiently and sustainably
Voluntary	Advisory services	Support from governments in the form of information, advice, and consultancy about improved wastewater treatment or solid waste management
	Voluntary agreements on wastewater treatment	Non-legally binding agreements negotiated on a case-by-case basis between wastewater treatment operators and a public authority to improve wastewater treatment practices
	Waste collection/ take-back schemes	Voluntary schemes designed to collect and appropriately dispose of unused pharmaceuticals, thereby reducing their release to water bodies.

Conclusion

Pharmaceutical wastewater has some characteristics such as poor biodegradability and high concentration due to the complexity of pharmaceutical processes. Treatment of pharmaceutical wastewater is particularly important because of these properties. There are many different types of advanced treatment, and each one has unique characteristics. The effluent quality of pharmaceutical wastewater can be effectively improved by the treatment technique mentioned above.

Pharmaceutical companies are subject to stringent rules from various nations' food and drug regulatory bodies, and they must maintain acceptable water quality standards for use, discharge, or reuse elsewhere in the plant. Huge quantities of ultrapure water are required with regulatory limitations on the limit or even presence of certain waste pollutants. Water released into towns or other waste streams may likewise be subject to volume restrictions. The main ways that pharmaceuticals leave the manufacturing facilities and end up in the environment are through usage and improper disposal. The majority of the medicines in the environment come from wastewater produced by fermentation and chemical synthesis processes at various production sites. During production, purification, cleaning, washing, and maintenance, these facilities produce a lot of waste.

References

[1] Ahmed, J., Thakur, A., & Goyal, A. (2021). "Industrial Wastewater and Its Toxic Effects." in *Biological Treatment of Industrial Wastewater*, edited by Maulin P. Shah. 1-14. https://doi.org/10.1039/9781839165399-00001.

[2] Kummerer, K. (2010). "Pharmaceuticals in the Environment." *Annu Rev Environ Reseor.* 35: 57-75. doi:10.1146/annurev-environ-052809-161223.

[3] Ram, S. K., Panidepu, H., Cheernam, V., Tyagi, R. D. (2020). "Pharmaceutical metabolites and their by-products in hospital wastewater." In *Current Developments in Biotechnology and Engineering*. Edited by Tyagi, R. D., Sellamuthu, B., Tiwari, B., Yan, S., Drogui, P. 43-78. https://doi.org/10.1016/B978-0-12-819722-6.00002-x.

[4] Al-Kaf, A. G., Naji, K. M., Abdullah, Q. Y. M., & Edrees, W. H. A. (2017). "Occurrence of Paracetamol in Aquatic Environments and Transformation by Micro organisms: A Review." *Chronicles of Pharmaceutical Science*, 1, 6: 341-355. https://www.scientiaricerca.com/srcops/SRCOPS-01-00033.php.

[5] Schacht, K., Chen, Y., Tarchitzky, J., & Marschner, B. (2016). "The use of treated wastewater for irrigation as a component of integrated water resources management:

reducing environmental implications on soil and groundwater by evaluating site-specific soil sensitivities." In *Integrated Water Resources Management: Concept, Research and Implementation.* Edited by Schacht, K., Chen, Y., Tarchitzky, J., & Marschner, B. 459-470. doi: 10.1007/978-3-319-25071-7_18.

[6] Victor, R., Kotter, R., O'Brien, G., Mitropoulos, M., & Panayi, G. (2008). "WHO Guidelines for the Safe Use of Wastewater, Excreta and Greywater." *Intern J. Envir Stud.,* 65, 1: 157-176. doi: 10.1080/00207230701846598.

[7] Maqbool, F., Mostafalou, S., Bahadar, H., & Abdollahi, M. (2016) "Review of endocrine disorders associated with environmental toxicants and possible involved mechanisms." *Life Sci.,* 145: 265–73. doi: 10.1016/j.lfs.2015.10.022.

[8] Nikolaou, A., Meric, S., & Fatta, D. (2007) "Occurrence patterns of pharmaceuticals in water and wastewater environments." *Anal Bioanal Chem.,* 387, 4: 1225–34. doi: 10.1007/s00216-006-1035-8.

[9] Grossberger, A., Hadar, Y., Borch, T., & Chefetz, B. (2013). "Biodegradability of pharmaceutical compounds in agricultural soils irrigated with treated wastewater." *Environ Pollut.,* 185: 168–177. doi: 10.1016/j.envpol.2013.10.038.

[10] Qishlaqi, A., Moore, F., & Forghani, G. "Impact of untreated wastewater irrigation on soils and crops in Shiraz suburban area, SW Iran." *Environ Monit Assess.,* 141,1–3: 257–73. doi: 10.1007/s10661-007-9893-x.

[11] Hani, A., & Pazira, E. (2011) "Heavy metals assessment and identification of their sources in agricultural soils of Southern Tehran, Iran." *Environ Monit Assess.,* 176, 1–4: 677–91. doi: 10.1007/s10661-010-1612-3.

[12] Behbahaninia, A., Mirbagheri, S. A. (2008). "Investigation of heavy metals uptake by vegetable crops from metal contaminated soil." *World Acad Sci Eng Technol,* 43: 56-68. https://www.scirp.org/(S(lz5mqp453edsnp55rrgjct55))/reference/references papers.aspx?referenceid=878496.

[13] Hani, A., Pazira, E., Manshouri, M., Kafaky, S. B., & Tali, M. G. (2010). "Spatial distribution and mapping of risk elements pollution in agricultural soils of southern Tehran, Iran." *Plant Soil Environ.,* 56, 6: 288–96. https://agris.fao.org/agris-search/search.do;jsessionid=93252CBDEA183EDBC176E13C04987EB8?request_locale=en&recordID=CZ2010000762&sourceQuery=&query=&sortField=&sortOrder=&agrovocString=&advQuery=&centerString=&enableField=.

[14] Bigdeli, M., & Seilsepour, M. (2008). "Investigation of metals accumulation in some vegetables irrigated with waste water in Shahre Rey-Iran and toxicological implications." *Am Eurasian J Agric Environ Sci.,* 4, 1: 86–92. https://www.sid.ir/en/Journal/ViewPaper.aspx?ID=449469.

[15] Nasr, M. M., & Yazdanbakhsh, A. (2008) "Study on wastewater treatment systems in hospitals of Iran." *J Environ Health Sci Eng.,* 5, 3: 211-215. http://www.bioline.org.br/request?se08036.

[16] Alobaidy, A. H. M. J., Al-Sameraiy, M. A., Kadhem, A. J., & Majeed, A. A. (2010). "Evaluation of treated municipal wastewater quality for irrigation." *J Environ Prot.,* 1,03: 216. doi: 10.4236/jep.2010.13026.

[17] Sato, T., Qadir, M., Yamamoto, S., Endo, T., & Zahoor, A. (2010). "Global, regional, and country level need for data on wastewater generation, treatment, and use." *Agr Water Manage.,* 130: 1-13. doi: 10.1016/j.agwat.2013.08.007.

[18] Larsson, D. (2014). "Pollution from drug manufacturing: review and perspectives." *Philosophical transactions of the Royal Society B, Biological sciences,* 369: 1656. https://doi.org/10.1098/rstb.2013.0571.

[19] Barakat, M. (2011). "New trends in removing heavy metals from industrial wastewater." *Arabian J. Chem.,* 4: 361 —377. https://doi.org/10.1016/j.arabjc.2010.07.019.

[20] Eccles, H. (1999). "Treatment of metal-contaminated wastes: why select a biological process." *Trends Biotechnol., **17:** 462 —465.* doi:10.1016/S0167-7799(99)01381-5.

[21] Chen, G. (2004). "Electrochemicals technologies in wastewater treatment." *Sep. Purif. Technol., 38, 1: 11-41.* https://doi.org/10.1016/j.seppur.2003.10.006.

[22] Szpyrkowicz, L., Naumczyk, J., & Zilio-Grandi, F. (1995). "Electrochemical treatment of Tannery wastewater using Ti/Pt/Ir electrodes." *Water Res.,* 29: 517-524. https://doi.org/10.1016/0043-1354(94)00176-8.

[23] Chiang, L. C., Chang, J. E., & Wen, T. C. (1995). "Indirect oxidation effect in electrochemical oxidation treatment of landfill leachat." *Water Res.*, 29: 671-678. https://doi.org/10.1016/0043-1354(94)00146-X.

[24] Rajkumar, D., Palanivelu, K., & Mohan, N. (2001). "Electrochemical oxidation of resorcinol for wastewater treatment using Ti/TiO2−RuO2−IrO2 electrode." *J. Environ. Sci. Health, Part A: Toxic/Hazard. Subst. Environ. Eng.,* 36:1997-2010. doi:10.1081/ESE-100107443.

[25] Rajkumar, D., & Palanivelu, K. (2003). "Electrochemical degradation of cresols for wastewater treatment." *Ind. Eng. Chem. Res.*, 42: 1833-1839. https://doi.org/10.1021/ie.020759e.

[26] Gupta, G. K., Liu, H., & Shukla, P. (2019) "Pulp and paper industry–based pollutants, their health hazards and environmental risks." *Curr. Opin. Environ. Sci. Health,* 12: 48 —56. https://agris.fao.org/agris-search/search.do?recordID=US2020 00023389.

[27] Singh, R. P., Singh, P. K., Gupta, R., & Singh, R. L. (2018). "Treatment and Recycling of Wastewater from Textile Industry." In *Advances in Biological Treatment of Industrial Waste Water and Their Recycling for a Sustainable Future.* Edited by Singh, R. L., Singh, R. P. Springer, 225-266. https://link.springer.com/book/10.1007/978-981-13-1468-1.

[28] Wichelns, D., Drechsel, P., & Qadir, M. (2015). "Wastewater: Economic Asset in an Urbanizing World." In *Wastewater: Econimic Asset in an Urbanizing World,* Edited by Drechsel, P., Quadir, M., and Wichelns, D. 3-14. Berlin: Springer. https://link.springer.com/book/10.1007/978-94-017-9545-6.

[29] UBA. (2018). "Recommendations for reducing micropollutants in waters, Background- April 2018." *German Environment Agency, Dessau-Roßlau.* https://www.umweltbundesamt.de/sites/default/files/medien/1410/publikationen/180709_uba_pos_mikroverunreinigung_en_bf.pdf.

[30] Daughton, C. G. (2014), "Eco-directed sustainable prescribing: Feasibility for reducing water contamination by drugs." *Science of the Total Environment,* 493: 392-404. https://doi.org/10.1016/j.scitotenv.2014.06.013.

Chapter 11

The Latest Trends in Bioremediation of Pharmaceutical Contaminants: Limitations and Future Prospects

Neha Agarwal[1,*], Rajeev Singh Yadav[2] and Vijendra Singh Solanki[3]

[1]Department of Chemistry, Navyug Kanya Mahavidyalaya, Lucknow, Uttar Pradesh, India

[2]Department of Chemistry, Ganna Utpadak P.G. College, Baheri, Bareilly, Uttar Pradesh, India

[3]Department of Chemistry, School of Liberal Arts and Sciences, Mody University of Science and Technology, Rajasthan, India

Abstract

A gargantuan acceleration in the use of pharmaceuticals in the last few decades has raised concerns over their occurrence in the soils and wastewater posing an imminent problem to the general public's health and the environment. Pharmaceuticals are one of the major contributors to the unsustainable environment because they are malignant and hardly biodegradable in nature. Numerous synthetic and semi-synthetic chemicals are being produced and consequently used globally for many different purposes, such as pharmaceuticals, pesticides, hydrocarbons, and many other toxic and persistent chemical species. The presence of the aforementioned substances not only in various environmental matrices (air, water, and soil) but also in biological systems has raised global scientific concerns regarding their potential toxicity towards non-target organisms, including humans. Considering the practical limitations of remediation techniques and measures in the treatment of recalcitrant

[*] Corresponding Author's Email: nehaagarwal4074@gmail.com.

In: Pharmaceuticals: Boon or Bane
Editor: Neha Agarwal
ISBN: 979-8-88697-487-4

xenobiotics, various bioremediation measures are currently being investigated and proposed for the removal of pharmaceutical contaminants from the environment. Several biological species like bacteria, fungi, and plant species have shown promising bioremediation potential for the removal of pharmaceutical waste from the environment. This article is intended to give an overview of the bioremediation techniques currently being employed or investigated and their latest trends to remove the pharmaceutical pollutants from the environment.

Keywords: biodegradable, contaminants, pharmaceuticals, bioremediation

1. Introduction

For the last many years, there has been a considerable jump in pharmaceutical industries output, which has registered, high economic growth on one hand but also given rise to environmental pollution on a massive scale. Untreated industrial waste water released from pharmaceutical industries poses severe environmental threats and hence generates a need for the assessment and characterization of discharged wastewater as per the norms of the regulatory authorities. Environmental challenges are a major problem in today's times [1, 2]. The waste water released from industries is quite hazardous to natural aquatic ecosystems [3, 4]. Expeditious industrialization worldwide has led to an increasing interrelationship between the environment and pollution indirectly. It is found that most of the rivers in the vicinity of industries are polluted [5].

These pollutants are a global problem and are not limited to just one region but they have been detected in waters around the world. Approximately 700 pharmaceuticals have been detected in aquatic systems in 71 countries [6]. Pharmaceutical compounds reach the environment as byproducts or metabolites due to human, veterinary, and industrial usage and end up in landfill leachates through sewage systems. These compounds enter the environment as byproducts of human and veterinary use through manufacturing waste, human excrement into septic tanks/sewage systems, animal excrement on soil combined with surface or agricultural runoff, household and hospital solid waste that end up in landfill leachates, and disposal of unused or expired medicine through sewage systems and landfills [7]. Due to accumulation of these pollutants, the global environment is under great stress and the problems are compounded by drastic changes that have been taking place in the lifestyles and habits of people. The major environmental threats include the release of organic aqueous waste (drugs and pesticides), organic liquids (solvents from dry-cleaning), oils (lubricating, automotive, hydraulic, and fuel oils), and organic sludge/solids (painting operations, tars from dyestuffs intermediates). During the past few decades, the pharmaceutical industries have registered a quantum jump worldwide. The pharmaceutical industry includes raw materials, antibiotics, a variety of medicines, and cosmetic products, which in turn generate a significant quantity of effluents containing constituents harmful to human and aquatic life [8, 9]. Because of the recalcitrant nature of these molecules, they survived through the wastewater treatment process and were finally discharged into the environment. Though the volume of untreated pharmaceutical industry wastewater is small, it contains a high concentration level of pollutants because of the presence of non-biodegradable organic remains (such as antibiotics, non-prescribed drugs, animal and plant steroids, reproductive hormones, beta-blockers, anti-inflammatory, analgesics, lipid regulators, antidepressants, personal care products, detergent metabolites, flame retardants, and other extensively used chemicals, i.e., spent solvents, reaction residues, used filter media, etc.,) heavy metals (such as lead, mercury, cadmium, nickel, chromium) and other pollutants [10, 11]. The presence of the aforementioned xenobiotic substances not only in various environmental matrices but also in biological systems as well as in several compartments of raw or processed food has raised global scientific concerns regarding their potential toxicity towards non-target organisms including humans. Predicted impacts of wastewater on the flora and fauna vary widely due to the wide variations in the characteristics of the wastewater. The impact of pharmaceutical chemicals on public health and environment is becoming increasingly important, not only due to their acute toxicity, but also to their genotoxic and mutagenic effects.

For the treatment of such non-biodegradable compounds, various Advanced Oxidation Processes (AOPs) are used in urban and industrial waste water treatment, which is undoubtedly an effective option for the treatment of waste water [12]. By applying the chemical treatments using AOPs; the contaminants in the wastewater can be transformed into more innocuous compounds that may also lead to complete mineralization of waste water [13]. The application of AOPs based on the available studies of the removal of pharmaceutical byproducts and metabolites present in wastewater which are not eliminated by other technologies can be analyzed. During the oxidation of wastewater, transformation products of the pharmaceuticals can be formed, which may be more biologically potent, less biodegradable, or more toxic compared to the parent compounds [14]. A more environment friendly approach that is used for the remediation of waste pollutants from the environment is bioremediation. These techniques are now extensively used and growing at an exponential rate in recent times because polluted sites have been successfully remediated using microbial processes. In the past few years, there have been developments in bioremediation techniques to successfully restore the environment in its original form. Though many physicochemical remedies are available for the degradation of pharmaceutical pollutants, the most promising solution for emerging pharmaceutical and other contaminants is bioremediation and natural reduction by the use of microbes. A variety of microbes, including aerobic and anaerobic bacteria and fungi, are involved in these bioremediation processes such as degradation, eradication, immobilization and detoxification of chemical wastes and organic pollutants from the environment by the action of microorganisms.

2. Treatment Approaches Used in Pharmaceutical Industries

2.1. Physicochemical Processes

Pharmaceutical industries employ a wide array of treatment methods for degradation of wastewater such as neutralization/pH adjustment, oxidation, sand filtration, ozone use, Fenton's method, coagulation/flocculation, electrocoagulation, photo-electrocoagulation, peroxi-electrocoagulation, peroxi-photo electrocoagulation, sedimentation, membrane separation, UV irradiation, adsorption, chlorination, distillation, solar photo-Fenton, reverse osmosis, bacterial treatments, fungal treatment, algal treatment, phyto-remediation and methods using membrane bioreactor (MBR), anaerobic fixed film reactor (AFFR), aerobic sequencing batch reactor (ASBR), membrane-aerated biofilm reactor (MABFR) and activated sludge [2, 9, 15, 16].

Few research utilized ozone for treatment of pharmaceutical wastewater, which required approximately 33 mg/l ozone under acidic conditions, whereas under alkaline conditions, 30 mg/l ozone was needed. It was reported that ozonation can improve biodegradability of wastewater at alkaline pH and higher treatment time favoured the enhanced biodegradability of wastewater [17]. Another group of researchers used electrocoagulation, photo electrocoagulation, peroxi-electrocoagulation and peroxi-photo electro-coagulation processes for the removal of chemical oxygen demand (COD) from pharmaceutical wastewater originating from Osvah Pharmaceutical Company [18]. To a certain extent, these pollutants can be removed from the environment via physical methods, biodegradation, and chemical advanced oxidation processes [19]. Among these technologies, ultraviolet (UV) irradiation and UV-based AOPs have received considerable attention in the past decade. But the contaminants cannot be eradicated by conventional wastewater treatment; they reach water bodies through many routes. Most active pharmaceutical ingredients (APIs) remain in liquid form and, to a certain extent, pass through a wastewater treatment plant. By using special techniques like membrane filtration, activated charcoal, or ozonation; the cleaning capacity of wastewater treatment plants can be improved. Having reached the water, pharmaceuticals and their metabolites may also get into groundwater, though in much lower concentrations. Traces of pharmaceuticals have also been detected in drinking water [20, 21]. Adsorption has been widely used for the treatment of wastewater, but the major drawback in this process is that the contaminant is transferred to the adsorbent only and there is a need of regeneration of adsorbent regularly which brings the additional high cost.

A number of researchers have reported that UV as part of AOPs for degrading micropollutants is more effective than direct UV photolysis treatment for many compounds [22, 23, 24]. Hydrogen peroxide (H_2O_2) which is one of the most powerful oxidizing agents is stronger than chlorine and chlorine dioxide. Irradiation with UV light, H_2O_2 can be converted into hydroxyl radical (•OH), which possesses a high oxidation potential (2.8 V) is also a potential technique for the degradation of pollutants. The formation of •OH with high redox potential in the UV/H_2O_2 process has led to the application of UV/H_2O_2 as a promising process for both water disinfection and oxidation of unwanted contaminants [25]. UV-based AOPs can be facilitated by processes other than UV/H_2O_2 or UV/TiO_2. The processes reported for the removal of micropollutants include UV/Fenton, UV/O_3 UV/O_3/H_2O_2, UV/H_2O_2/TiO_2, UV/O_3/TiO_2, UV/HOCl, UV/ClO_2, and ultrasound/UV techniques. The mechanisms of UV/O_3 and UV/ultrasound both involve the formation of •OH via photolysis. O_3 is a strong oxidizing agent and reacts with organic compounds through a direct pathway via molecular ozone and a radical pathway by means of hydroxyl radicals. The UV radiation when combined with O_3 may be a more effective advanced oxidation technique than using O_3 alone for certain target compounds due to the formation of additional H_2O_2 and •OH radical generation via photolysis.

According to a study, pharmaceutical industrial wastewater contained nalidixic acid, which is an antibiotic pertaining to the quinolone group; cannot be easily biodegraded [26]. The biodegradability of nalidixic acid was achieved by the chemical oxidation process followed by biological treatment. The chemical oxidation process enhances the biodegradability, followed by biological treatment using an immobilized biomass reactor (IBR). The combined efficiency (physicochemical treatment along with bioremediation) of treatment was above 95%. Another study was done on the combined chemical treatment of pharmaceutical effluents from medical ointment production. A Fenton-like system was used in combination with lime coagulation where all effluent samples were subjected to pretreatment by an adsorption/flocculation/filtration process. Under the most favorable treatment conditions, COD and BOD (biological oxygen demand) removal of 87–96 and 79–95% was achieved, respectively [1]. Figure one shows various treatment techniques for the removal of contaminants/xenobiotics from the environment.

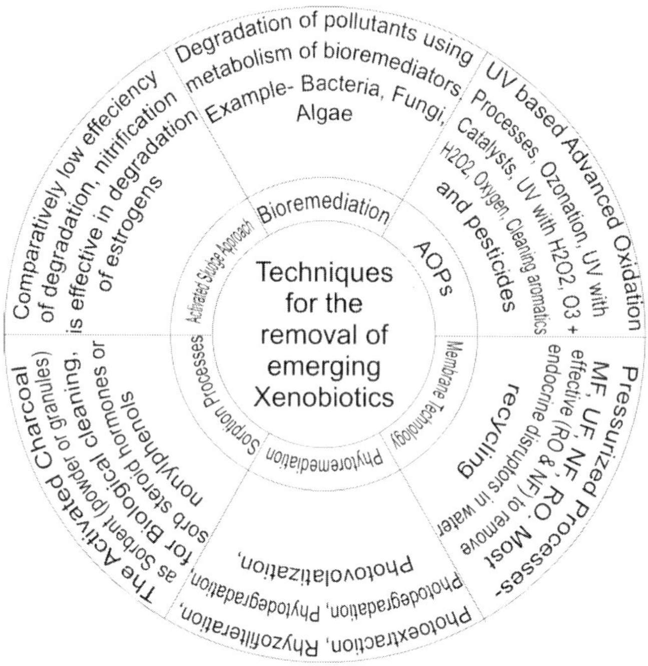

Figure 1. Various treatment techniques applied for the removal of xenobiotics from the environment. Information collected from [27, 28, 29].

The application of combined treatment of Fenton-like treatment and lime coagulation improved the quality of effluents, met the requirements for wastewater discharge to sewage and also increased the biodegradability of pharmaceutical effluents. Though UV-based AOPs have been proven to be efficient for the removal of a variety of micropollutants in aquatic environment but there are certain drawbacks associated with these treatments like high energy consumption, very high operational costs, unavailability of chemicals, and being non-selective in nature these result in the formation of end products with more toxicity than the parent pollutant compound. Though for full scale treatment and reuse of wastewater; chemical membrane technologies such as reverse osmosis, nanofiltration and ultrafiltration have been used. But in addition to high cost these methods have many operational difficulties. Therefore, for complete removal of contaminants from the environment physical methods alone may not be suitable [30, 31, 32].

2.2. Bioremediation Techniques

2.2.1. What Is Bioremediation?

It is a process where biological microorganisms are used to degrade environmental pollutants through metabolic processes. The term "biological" implies microscopic organisms such as bacteria, fungi, algae, and the remediation implies the treatment of the problem (organic pollutants). It may also be defined as a *"treatment technique done by naturally occurring organisms to break down toxic waste materials from the environment and to convert them into non-toxic materials"* may be employed in order to attack a different variety of contaminants, such as pharmaceuticals, chlorinated pesticides that are degraded by bacteria, or a more general approach may be taken, such as oil spills that are broken down using multiple techniques, including the addition of biosurfactant to facilitate decomposition of crude oil by bacteria [33]. Bioremediation may be either aerobic or an-aerobic[34, 35]. Sometimes; sequential anaerobic–aerobic bioremediation processes are also adopted to remediate contaminated sites [36]. Bioremediation may be shown schematically for the following process. On the basis of removal of waste, bioremediation can be classified into in situ and ex situ bioremediation techniques as mentioned in Figure 2.

Assessment and characterization of wastewater are very important to evaluate the quality of wastewater. In India, the Central Pollution Control Board (CPCB) provides standards with their limiting concentrations for the discharge of environmental pollutants from the pharmaceutical industry.

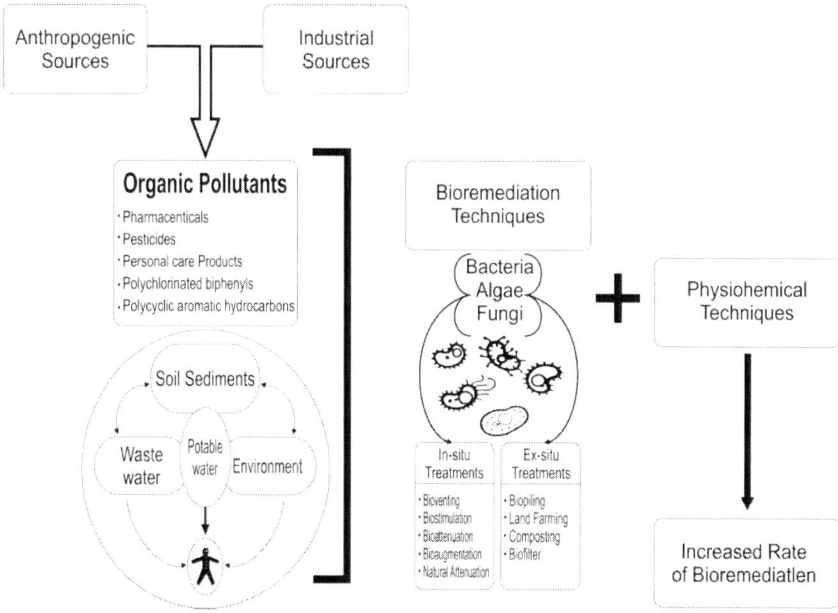

Figure 2. An overview of the microbial bioremediation techniques for pollutants.

Nowadays, pharmaceutical companies employ a variety of treatment methods, which include primary: chemical and physicochemical, secondary: biological process tertiary: advanced oxidation processes [37]. Among all other techniques, bioremediation is one of the latest and most widely used techniques for the treatment of pharmaceutical industry wastewater. Scanty of work has been done on characterization assisted bioremediation methods in India and available reports are also limited. Thus, identifying the need to explore the area in light of the gaps and lack of expensive work done, the present chapter gives a detailed account of available work in this field.

2.2.2. Microbes Involved in Bioremediation

Microorganisms are an integral part of the nutritional food chain and are indispensable to maintain the biological balance in life. The process of bioremediation involves the degradation of toxic materials with the help of microorganisms like bacteria, algae, fungi, and yeast. The microbes have the ability to grow and survive in lower and upper extremes of temperatures in the presence of hazardous compounds. The microbial adaptability and biological systems make them highly suitable for the remedial degradation process [38]. Though every microorganism has the ability to eradicate pollutants, only a few particular or engineered microorganisms are used broadly to eradicate pollutants efficiently, such as *Achromobactor*, *Alcaligenes*, *Bacillus*, *Pseudomonas*, *Corynebacterium*, *Flavobacterium*, *Mycobacterium*, *Nitrosomonas*, *Xanthobactor*, *Escherichia*, *Klebsiella*, and *Enterobacter* etc. [39]. Among them, Bacillus species have been extensively employed in remediation of organic pollutants. Microorganisms have a vital role in the degradation of xenobiotics and in maintaining the steady-state concentration of pollutants in the environment [40]. Microbial transformations result in residues that are less toxic and more stable than the parent compound. Biological treatment of pharmaceutical industrial effluents in wastewater can be achieved by aerobic and anaerobic treatment [36]. The biodegradation of the product depends on a number of factors, such as stereochemistry of the compound, toxicity of the pollutant compound, its concentration, the efficiency of the microbial strain, degradation conditions, retention time, and the presence of other compounds and their concentration [41]. Aerobic bacteria (*Pseudomonas*, *Nacordia*, *Flavobacterium*, *Mycobacterium*, *Acinetobactor*) can degrade complex xenobiotic compounds such as pesticides, hydrocarbons, and polyaromatic compounds, and many of them use these contaminants as a source of carbon and energy. Anaerobic bacteria have also been used for bioremediation of polychlorinated phenols, toxic aromatic compounds, and other toxic materials into non-toxic or less toxic forms.

2.2.3. Bioremediation of Pharmaceuticals and Other Effluents

The conventional chemical treatment methods of pharmaceutical industrial effluents have not been found completely successful in overcoming the toxic pollution load of global effluents and sometimes they also contribute to another type of complex by-product, which is more difficult to treat and further pollutes the soil or aquatic sources. For example, the chlorinated phenols are more toxic than the unsubstituted phenols because the toxicity of phenols increases with the degree of chlorination [42]. Many xenobiotics persist in the environment and do not undergo biological transformation due to their recalcitrant nature. Physicochemical purification methods utilize expensive chemicals and treatment units, which are difficult to manage at the industrial unit level [43].

Compared to other technologies, bioremediation is an eco-friendly process that utilizes microorganisms for the removal of organic pollutants from the natural environment (soil and water etc.). The treatment method is not only composed of a single strain but also employs diverse, synergistic or antagonist communities in the natural environment. Among microbial communities, bacteria, algae, and fungi have the capability to bio-transform persistent xenobiotics, rendering them nontoxic. Fungi and bacteria are recognized to produce different enzymes which act on a broad array of organic contaminants and help in the degradation of organic pollutants. Bioremediation technologies are coming into extensive use and continuation today at an exponential rate. Remediation of polluted sites has proven quite effective and effaceable due to its eco-friendly nature using microbial processes. Economic viability and environment friendly nature are major advantages of bioremediation compared to other methods of remediation. Indigenous microorganisms present in contaminated environments are the key to solving most of the challenges associated with biodegradation of pollutants [44]. Bioremediation involves the reduction, detoxification, mineralization and transformation of highly toxic xenobiotics into less toxic or simpler materials. The removal efficiency mainly depends upon the nature of the pollutant (pesticides, chlorinated compounds, pharmaceuticals, personal care products, heavy metals, hydrocarbons, dyes, plastics, sludge, etc.). Bioremediation is done to eradicate, immobilize, detoxify the diverse organic wastes and hazardous chemicals from the environment and ecosystem through the all-inclusive action of microorganisms.

2.2.4. Anaerobic Treatment

Previously, anaerobic bioremediation technology was used for treating wastewater from different industries, such as paper and pulp, distilleries, tanneries, textile and food processing, ranging from high-strength waste to low-strength waste. For such purposes various reactor configurations such as anaerobic contact reactor (ACR), fluidized bed reactor (FBR), upflow anaerobic sludge blanket reactor (UASB), and anaerobic fixed film reactor (AFFR) have been developed to treat wastewaters from different industries. Each of these configurations has different operational constraints, even though they are being used extensively [45]. With the help of microorganisms, the high organic content in industrial wastewater decomposes into methane and CO_2 in an anaerobic treatment process. Anaerobic treatment of industrial wastewater shows many interesting advantages such as production of very little sludge, lower energy requirements, operation at high organic loading rate, need of low nutrient amount, and production of biogas, which can be used to generate energy in this treatment process [46].

2.2.5. Aerobic Treatment

In this degradation method, generally aerobic sequencing batch reactors (ASBR) and activated sludge processes have been used for the treatment of industrial waste. This process consists of a primary settling tank, an intermediate retention trough, two storage tanks and an aerobic tank. Some research has reported the aerobic biological treatment of pharmaceutical wastewater [47]. Another research reported the use of activated sludge systems for treatment of wastewater of the herbal pharmaceutical industry and achieved a 68.98–91.02% reduction in COD, 76.90–97.26% reduction in BOD, and a 74.61–95.54% reduction in suspended solids [48]. Aerobic and anaerobic biological treatments employed for wastewater treatment are not able to remove the high concentration of organic pollutants completely. Therefore, there is a need for a handy, simple, cost-effective, and eco-friendly method to effectively treat the wastewater [49].

2.2.6. Bacterial Degradation

These are the classes of microorganisms that are involved in the degradation of organic pollutants from polluted sites and are vital in sustainable development of the environment and geochemical cycle. Several studies have been done on the degradation of environmental xenobiotics by different bacteria exclusively on hydrocarbons, and these bacteria are known as hydrocarbon-degrading bacteria. Bacterial culture for bioremediation of industrial wastewater has also been used and demonstrated in many studies. Some bacterial strains like *Pseudomonas*, *Enterobactor*, Streptomonas, *Aeromonas*, *Acinetobacter*, and *Klebsiella* show up to a 44% COD reduction [50]. Another group of researchers isolated 15 rhizosphere bacteria which showed 76% color reduction and 85–86% BOD and COD reduction within 30 days [51].

These microorganisms utilizes nitrate (denitrifying organisms), sulfate (sulfate reducers), Fe (III) (ferric- iron reducer), CO_2 (methanogens), and many other electron acceptors (chlorate, Mn, etc.) depending upon their electron – acceptor availability and competition of electron donating ability in different microorganisms [52]. The bacterial community is required to provide all metabolic capabilities for complete mineralization of these toxic contaminants, which is essential for degradation of organic pollutants [53]. Soil bacteria (especially *Pseudomonas*) predominantly in rhizospheric soil, have a special biodegradation potential for phenolics [54]. The bacterial capability to degrade phenol and chlorophenols has also been reported in another study done by Agarry and Soloman [55]. Another study has been done on degradation of pesticides by bacterial consortiums *(Acinetobacter* sp., *Pseudomonas putida, Bacillus* sp., *Pseudomonas aeruginosa, Citrobacter freundii, Stenotrophomonas* sp., *Flavobacterium* sp., *Proteus vulgaris, Pseudomonas* sp., *Acinetobacter* sp., *Klebsiella* sp., *and Proteus* sp.) isolated from contaminated sites. The result showed that the bacterial consortia were able to degrade pesticides successfully [56]. The control of pharmaceutical effluent parameters through bioremediation was studied by another group of researchers [57].

2.2.7. Fungal Degradation

Fungi are organisms that are ubiquitous and play an important and vital role in all ecosystems. These are microscopic eukaryotic organisms that have the capability to grow on different substrates and are competent at carrying out their functions almost for an indefinite period of time. These are the diverse groups of organisms such as yeast, molds, and filamentous fungi that have the capability to remediate waste from various industrial wastewaters. There are a number of fungal strains that play an important role in the treatment of industrial wastewater, but have certain limitations due to the presence of a long growth cycle and the formation of spores. A comparative study was done on the fungal treatment with the activated sludge treatment for the of pharmaceutical industrial wastewater where a fungal strain was used, through which 91% COD reduction was achieved as compared to activated sludge, which reduced 78% COD [2].

Fungi are well known to have an efficient capability of degradation where they destroy different types of materials like wood, textiles, plastic, leather, and paper. Fungi also help in the degradation of dyes, pesticides, polychlorinated biphenyls, chlorinated and phenolic compounds, hydrocarbons by the use of different enzymes produced by them such as laccases, manganese peroxidase (MnP) and lignin peroxidases (LiP) [58]. Many fungi (*Fusarium oxysporum, Trichoderma viride and Phanerochaete chrysosporium*) are known to have capability to degrade DDT. Fungi use lignin peroxidases for the mineralization of DDT and in some fungal strains the rate of mineralization depends upon the source of carbon. The rate and extent of mineralization of DDT is lesser as compared to its analogues such as dicofol and methoxychlor.

2.2.8. Algal Degradation

Many microalgae and protozoa are present in the ecosystem which is involved in various activities, but only a few of them are reported in the degradation of hydrocarbons, pesticides, and persistent pollutants [59]. Many algal species (*Rhodococcus* genus, *Caepidiumantarticum*, *Desmarestia* sp., *Focus* sp. and *Ascophyllum nodosum*) are found to be effective in efficient degradation of hydrocarbons [60] and pesticides are degraded by many algae. In a research study an artificial microalgae-bacterial consortium was constructed for the degradation of crude oil and the results reported that hydrocarbons were efficiently degraded by this artificial consortium [61]. Algae that are capable of degrading petroleum hydrocarbons into their less toxic form have the capability of functioning with the help of certain enzymes produced by them. Many other algal species like *Rhodococcus* genus, *Caepidiumantarticum, Desmarestia* sp., *Focus* sp. *and Ascophyllum nodosum* are also found to be very effective in efficient degradation of hydrocarbons and pesticides [60].

3. Latest Trends and Advances

Recently, the processes of biodegradation and bioremediation have been revealed as the latest trends in the removal of pollutants from the environment. Scientists have shown the application of genetically modified microbes and metagenomics, microbial fuel cells, nano-materials, biofilms, and constructed wetlands for the degradation of persistent organic pollutants from different sources. A multidisciplinary involvement of genomics, microbiology, molecular biology, biochemistry, and bioinformatics is required to produce genetically engineered microorganisms (GEMs) to overcome various hurdles in the cleanup of contaminated sites. There are very few studies where GEMs have been applied and proven to be more efficient than natural MOs in the elimination of recalcitrant compounds under natural conditions [62].

3.1. Genetically Modified Microorganism for Bioremediation

The main aim of biomolecular engineering in bioremediation is to enhance the biocatalytic capability of microbes to degrade xenobiotics. For bioremediation purposes, the first test release of genetically engineered microorganisms was commenced by the US Environmental Protection Agency (EPA) in 1996. It is a relatively new field that exploits engineering biomolecules and biomolecular processes for the development of genetically engineered enzymes or microbes for bioremediation purposes.

Out of the large number of enzymes that help in the degradation of pollutants, Cytochrome P450 monooxygenase (CYPs) is one of the largest groups that degrade these toxic pollutants by forming catechols, which are further degraded by catechol dioxygenase to form various products in microorganisms. The wild type cytochrome P450 enzyme from Bacillus megaterium has been shown to have an inherently low activity towards naphthalene, fluorine, acenaphthene, acenaphthylene, and 9-methylanthracene. Therefore, several mutations have been done based on their crystal structure to obtain a cytochrome P450 BM-3, which showed the highest activity towards degradation of pollutants and was found to have a great potential in degradation of them in the environment [63].

3.2. Microbial Fuel Cells

Microbial fuel cells (MFCs) are chemical systems that generate electricity from the biodegradation of organic materials with the help of a suitable microbial substrate. For decades, this technique has been used as a renewable energy source that requires more power than wastewater treatment. The first laboratory substrates used were mainly glucose, acetate, or other simple substrates to determine the behavior of electrode materials, membranes, as well as the reactor architecture or microbial activity. The studies pertaining to wastewater as a substrate began in 2004 only. Sufficient literature is available to study the extensive description of MFCs and their operational principles, their electric performance, their ion and electron transport mechanisms etc. [64, 65, 66, and 67].

Essentially, wastewater is the most popular and suitable substrate for an MFC operation due to its high percentage in organic load and minimal or zero cost, especially the agro-food wastewater is very suitable due to its high degree of biodegradability [68, 69]. Although MFCs not yet have economically viable prices per kW of installed power and with a high nominal power (equivalent to renewable power plants of the order of MW), much recent research on MFCs has, however, been progressing rapidly and continuously in the coming time.

3.3. Nanomaterials in Degradation of Pollutants

Nano-materials (also known as nano-particles) are the smallest structures, having a size of less than 100 nm. A variety of efficient, cost-effective and eco-friendly nanomaterials have been developed which have unique potential for wastewater treatment (industrial effluent, surface water, ground water and drinking water). Nanomaterials have attracted a lot of interest in terms of environmental remediation. Due to their huge surface area and related high reactivity, nanomaterials perform better in environmental clean-up than other conventional approaches. They can also be modified for specific uses to provide novel features. Due to the large surface-area-to-volume ratio and the presence of a larger number of reactive sites, nanomaterials can be extremely reactive and useful. These characteristics allow for higher interaction with contaminants, which in turn leads to a quick and efficient reduction of contaminant concentration.

Antibiotic activity has been shown by green synthesized particles (especially silver nanoparticles). They have been used for medical implantations such as silicone artificial heart valves, catheters, bone prosthetics, dental sets, and bandages [70]. Another class of nanomaterials is used as nano-catalysts (metal oxides and semiconductors) have gained significant popularity in developing technologies for wastewater treatment. Different types of nano-catalysts such as electrocatalysts, Fenton based catalysts (improve chemical oxidation of organic pollutants) and catalysts having antimicrobial properties have been employed for the degradation of pollutants [71]. Another important class of nano-materials is nano-membranes which is a pressure-driven treatment technique used in improving the quality of wastewater. Due to its small pore size, low cost, high efficiency, and user-friendly nature, nano-filtration is extensively applied for treatment of wastewater in industries among various types of membrane filtration techniques available.

3.4. Importance of Metagenomics and Metaproteomics in Bioremediation

It is difficult to cultivate the microbial population that establish a complex ecological niche at the site of contamination in the environment. Therefore, it is very important to understand the detailed structure of the composition of communities and their role in bioremediation of various toxic compounds.

This can be done by direct DNA isolation from the bacterial community, which will help us to understand the degradation mechanism of a specific ecosystem. Exposure to organic pollutants has detrimental effects on life present on earth through diverse mechanisms that lead to the formation of toxic reactive intermediates and metabolites and also disrupt endogenous metabolism due to changes in their environment. Metagenomics and metaproteomics may create a novel approach for studying site-specific microorganisms during the active mineralization process. Earlier, metagenomics studies were done to identify the bacterial community by cultivation [72]. Recent investigations have been conducted to identify the biodegradation potential pathways of persistent pesticides in freshwater and marine sediments. It was reported that *Plesiocystis*, *Anaerolinea*, *Jannaschia*, and *Mycobacterium* were found to be an effective population present in all sediments, which helps in biodegradation [73].

4. Limitations Associated with the Degradation of Polluants

Along with many advantages, there are certain limitations to bioremediation. Depending upon the site; some contaminants may not be completely transformed to harmless products or the intermediate compound formed may be even more toxic than the parent compound and non-biodegradable. Heavy metals and other organic pollutants generally inhibit the activity of indigenous microorganisms. *In-situ* bioremediation requires microorganism acclimatization, which may not be developed for recalcitrant compounds because many of these pollutants are recalcitrant in nature. Microbial technology is basically restricted to those compounds that are biodegradable, but there are many such compounds that cannot be completely or rapidly degraded. The degradative pathway of many pollutants is not completely known, and many industrial pollutants are not yet identified. According to some studies, fungi do not use persistent hydrocarbon pollutants as a carbon and energy source; thus, additional carbon sources are supplemented for their metabolism.

5. Future Prospects

Diversified research has been going on continuously to find an effective and eco-friendly solution to increasing environmental pollution. Long-term measures for both human and veterinary pharmaceuticals are being considered for the development of new pharmaceuticals as well as their redesign by applying a "benign by design" approach for improved diagnostics. There is a great focus on the isolation, characterization and purification of novel enzymes and their roles in the bioremediation of organic pollutants. Although it is a time-consuming process, it can be overcome by using metagenomics and metaproteomics approaches. In addition bioremediation processes are technological tools that promise positive results for the future because they do not produce waste by-products, have the potential to be economically viable, and, in combination with the different physicochemical and advanced oxidation processes, offer better results for the treatment of pharmaceutical industrial wastewater and thus help in utilizing the available water resources in a sustainable manner. Most likely, it is to become one of the best technologies used to clean up and protect our environment when used with the latest advancements.

Conclusion

Environmental problems caused by a variety of pharmaceuticals, pesticides, dyes, and other industrial effluents are mainly due to the accumulation of pollutants and other fragmented compounds, which in turn form into other substitutes (natural or manmade), finally forming a xenobiotic. There is an urgent need to degrade these xenobiotic compounds in an eco-friendly manner. Various techniques like microbial remediation, bioremediation, and phytoremediation and their subtypes have been discussed. Each has its own method of degrading these xenobiotics, as well as the consequences of fragmentations and bioaccumulations in the environment. Bioremediation depends upon many factors, but it is not limited to the type and concentration of pollutants. Geological characteristics of polluted sites (soil, pollutant type and depth, human habitation sites) and the performance of bioremediation techniques should be integrated in determining the most appropriate technique to successfully treat the polluted site. A novel equipment-based technique which is rapid but also has a negative impact on the environment is a solar-driven technique. Phytoremediation is restricted to particular sites containing contaminants. Although slow, on the whole, microbial bioremediation can cover a wide range of recalcitrant degradation and is known to be a better choice because of its nature of degradation.

References

[1] Kaushik, K.A., Dalal, S.J., &Panwar, S. (2012). "Impact of Industrialization on culture of Uttarakhand and its role in career enhancement." *VRSD Int J Bus Manage Res.*, 2, 4:123-132. https://scholar.google.com/citations?view_op=view_citation&hl=en&user=lSBkRrMAAAAJ&citation_for_view=lSBkRrMAAAAJ:9yKSN-GCB0IC.

[2] Spina, F., Anastasi, A., Prigione, V., & Varese, G.C. (2012). " Biological treatment of industrial wastewaters: a fungal approach." *ChemEng Trans.*, 27: 175-180. https://doi.org/10.3303/CET1227030.

[3] Deepali, J. N. (2012). " Study of ground water quality in and around SIDCUL industrial area, Haridwar, Uttarakhand, India." *J ApplTechnol Environ Sanitation,* 2,2:129-134. https://journaldatabase.info/articles/study_ground_water_quality_around.html.

[4] Lokhande, R.S., Singare, P.U., & Pimple, D.S. (2011). "Study on Physico-Chemical Parameters of Waste Water Effluents from Taloja Industrial Area of Mumbai, India. *International Journal of Ecosystem*, 1, 1-9. https://doi.org/10.5923/j.ije.20110101.01.

[5] Modak, D.P., Singh, K.P., Ahmed, S., & Ray, P.K. (1990). "Trace metal ion in Ganga water system." *Chemosphere,* 21,1-2: 275-287. https://agris.fao.org/agris-search/search.do?recordID=GB9107738.

[6] Aus der Beek, T. Weber, F. A., Bergmann, A., Hickmann, S., Ebert, I., Hein, A., & Kuster, A. (2016). " Pharmaceutical in the Environment -Global Occurrences and Perspectives." *Environmental Toxicology and Chemistry,* 35, 4: 823-835. https://doi.org/10.1002/etc.3339.

[7] Agarwal, N. (2022). "Paracetamol-A Contaminant of high concern: Existence in Environment & Adverse Effects." *Pharmaceutical Drug Regulatory Affairs Journal*, 5, 1:1-8.doi: 10.23880/pdraj-16000128.

[8] Dixit, D., & Parmar, N. (2013). "Treatment of pharmaceutical waste water by electro-coagulation and natural coagulation process: review." *VSRD Int J TechnolBioTechnol Res.,* 4, 5: 79–88. https://scholar.google.com/citations?view_op=view_citation&hl=en&user=Cn-S-boAAAAJ&citation_for_view=Cn-S-boAAAAJ:d1gkVwhDpl0C.

[9] Chang, C.Y., Chang, J.S., Vigneswaran, S., & Kandasamy, J. (2008). "Pharmaceutical wastewater treatment by membrane bioreactor process—a case study in southern Taiwan." *Desalination,* 234:1–9. doi: 10.1016/j.desal.2007.09.109.

[10] Vuppala, N.V.S., Suneetha, C., & Saritha, V. (2012) "Study on treatment process of effluent in Bulk drug industry." *Int J Res Pharm Biomed Sci.*, 3,3: 1095–1102. https://www.researchgate.net/publication/291802549_Study_on_treatment_process_of_effluent_in_Bulk_drug_industry.

[11] Ramola, B., & Singh, A. (2013). "Heavy metal concentrations in pharmaceutical effluents of Industrial Area of Dehradun (Uttarakhand), India." *Int J Environ Sci Res.*, 2, 2:140–145. doi: 10.4172/2161-0525.1000173.

[12] Boczkaj, G., &Fernandes, A. (2017). "Wastewater treatment by means of advanced oxidation processes at basic pH conditions: A review." *Chem. Eng. J.*, 320, 608–633. https://doi.org/10.1016/j.cej.2017.03.084.

[13] Poyatos, J.M., Muñio, M.M., Almecija, M.C., Torres, J.C., Hontoria, E., & Osorio, F. (2010) "Advanced oxidation processes for wastewater treatment: State of the art." *Water Air Soil Pollut.*, 205, 187–204. https://doi.org/10.1007/s11270-009-0065-1.

[14] Bletsou, A.A., Jeon, J., Hollender, J., Archontaki, E., & Thomaidis, N.S. (2015) "Targeted and non-targeted liquid chromatography-mass spectrometric workflows for identification of transformation products of emerging pollutants in the aquatic environment." *Tr AC Trends Anal. Chem.*, 66, 32–44. https://doi.org/10.1016/j.trac.2014.11.009.

[15] Patil, P.G., Kulkarni, G.S., Kore, S.V., & Kore, V.S. (2013). "Aerobic sequencing batch reactor for wastewater treatment: a review." *Int J Eng Res Technol,* 2,10: 534–550. doi: 10.17577/IJERTV2IS100087.

[16] Madukasi, E.I., Dai, X., He, C., & Zhou, J. (2010) "Potentials of phototrophic bacteria in treating pharmaceutical wastewater." *Int J Environ Sci Technol,* 7, 1: 165–174. https://doi.org/10.1007/BF03326128.

[17] Gome, A., & Upadhyay, K. (2013). "Biodegradability assessment of pharmaceutical wastewater treated by ozone." *Int Res J Environ Sci.,* 2,4: 21–25. http://www.isca.in/IJENS/Archive/v2/i4/5.ISCA-IRJEvS-2013-054.php.

[18] Farhadi, S., Aminzadeh, B., Torabian, A., Khatibikamal, V., & Fard, M.A. (2012) "Comparison of COD removal from pharmaceutical wastewater by electro-coagulation, photo electrocoagulation, peroxi-electrocoagulation and peroxi-photo electrocoagulation processes." *J Hazard Mater.,* 219–220: 35–42. doi: 10.1016/j.jhazmat.2012.03.013.

[19] Liu, Z. H., Kanjo, Y., & Mizutani, S. (2009). "Removal mechanisms for endocrine disrupting compounds (EDCs) in wastewater treatment – physical means, biodegradation, and chemical advanced oxidation: A review." *Science of the Total Environment,* 407,2: 731–748. https://doi.org/10.1016/j.scitotenv.2008.08.039.

[20] Kleywegt, S., Pileggi, V., Yang, P., Hao, C., Zhao, X., Rocks, C., Thach, S., Cheung, P., & Whiteland, B. (2011). "Pharmaceuticals, hormones and bisphenols A in untreated source and finished water in Ontario, Canada: Occurrence and Treatment Efficiency." *Science of the Total Environment,* 409, 8: 1481-1488. doi: 10.1016/j.scitotenv.2011.01.010.

[21] Togola, A., & Budzinski, H. (2008). "Multi-residue analysis of pharmaceutical compounds in aqueous samples." *Journal of Chromatography A.,* 1177, 1: 150-158. doi: 10.1016/j.chroma.2007.10.105.

[22] Chen, P. J., Linden, K. G., Hinton, D. E., Kashiwada, S., Rosenfeldt, E. J., & Kullman, S. W. (2006). "Biological assessment of bisphenol-A degradation in water following direct photolysis and UV advanced oxidation." *Chemosphere,* 65, 7: 1094–1102. doi: 10.1016/j.chemosphere.2006.04.048.

[23] Rosenfeldt, E. J., & Linden, K. G. (2004). "Degradation of endocrine disrupting chemicals bisphenol A, ethinyl estradiol, and estradiol during UV photolysis and advanced oxidation processes." *Environmental Science & Technology,* 38, 20: 5476–5483. https://doi.org/10.1021/es035413p.

[24] Feng, X. H., Ding, S. M., Tu, J. F., Wu, F., & Deng, N. S. (2005). "Degradation of estrone in aqueous solution by photo-Fenton system." *Science of the Total Environment,* 345, 1–3: 229–237. https://doi.org/10.1016/j.scitotenv.2004.11.008.

[25] Kruithof, J. C., Kamp, P. C., & Martijn, B. J. (2007). "UV/H_2O_2 treatment: A practical solution for organic contaminant control and primary disinfection." *Ozone Science & Engineering,* 29, 4: 273–280. doi: 10.1080/01919510701459311.

[26] Sirtori, C., Zapata, A., Oller, I., Gernjak, W., Aguera, A., & Malato, S. (2009). "Decontamination industrial pharmaceutical wastewater by combining solar photo-Fenton and biological treatment." *Water Res.,* 43: 661–668. doi: 10.1016/j.watres.2008.11.013.

[27] de Oliveira, J.F., Fia, R., Rodrigues, F.N., Fia, F.R.L., de Matos, M.P., Siniscalchi, L.A.B., & Sanson, A.L. (2020). "Quantification, removal and potential ecological risk of emerging contaminants in different organic loads of swine wastewater treated by integrated biological reactors." *Chemosphere,* 260: 127516. https://doi.org/10.1016/j.chemosphere.2020.127516.

[28] Lopera, A.E.C., Ruiz, S.G., & Alonso, J.M.Q. (2019). "Removal of emerging contaminants from wastewater using reverse osmosis for its subsequent reuse: Pilot plant." *J. Water Process. Eng.,* 29: 100800. https://doi.org/10.1016/j.jwpe.2019.100800.

[29] Mohapatra, D.P., & Kirpalani, D.M. (2019). "Advancement in treatment of wastewater: Fate of emerging Contaminants." *Can. J. Chem. Eng*., 97: 2621–2631. https://doi.org/10.1002/cjce.23533.

[30] Chakraborty, J., & Das, S. (2016). "Molecular perspectives and recent advances in microbial remediation of persistent organic pollutants." *Environmental Science and Pollution Research,* 23,17: 16883-16903. doi: 10.1007/s11356-016-6887-7.

[31] Krishnan, S., Rawindran, H., Sinnathambi, C., & Lim., J. (2017). "Comparison of various advanced oxidation processes used in remediation of industrial wastewater laden with recalcitrant pollutants." *IOP Conference Series: Materials Science and Engineering. IOP Publishing,*206, 1: 012089. doi: 10.1088/1757-899X/206/1/012089.

[32] Muruganandham, M., Suri, R., Jafari, S., Sillanpää, M., Lee, G. J., Wu, J., & Swaminathan, M. (2014). "Recent developments in homogeneous advanced oxidation processes for water and wastewater treatment." *International Journal of Photoenergy*, 2014: 1-21. https://doi.org/10.1155/2014/821674.

[33] Juwarkar, A.A. and Singh SK. 2010 "Microbe-assisted phytoremediation approach for ecological restoration of zinc mine spoil dump." *Int J Environ Poll*, 43:236–250. https://www.inderscienceonline.com/doi/abs/10.1504/IJEP.2010.035927.

[34] Wiegel, J., & Wu, Q. (2000) "Microbial reductive dehalogenation of polychlorinated biphenyls." *FEMS Microbiol Ecol.,* 32: 1–15. doi: 10.1111/j.1574-6941.2000.tb00693.x.

[35] Komancova, M., Jurcova, I., Kochankova, L., & Burkhard, J. (2003) "Metabolic pathways of polychlorinated biphenyls degradation by *Pseudomonas* sp." *Chemosphere,* 50: 537–543. https://doi.org/10.1016/S0045-6535(02)00374-0.

[36] Deegan, A.M., Shaik, B., Nolan, K., Urell, K., Oelgemöller, M., Tobin, J., & Morrissey, A. (2011). "Treatment options for wastewater effluents from

pharmaceutical companies." *Int J Environ Sci Technol.,* 8, 3: 649–666. https://www.academia.edu/40908645/Treatment_options_for_wastewater_effluents_from_pharmaceutical_companies.

[37] Wei, X., Li, B., Zhao, S., Wang, L., Zhang, H., Li, C., & Wang, S. (2012) "Mixed pharmaceutical wastewater treatment by integrated membrane aerated biofilm reactor (MABR) system—a pilot-scale study." *Bioresour Technol,* 122: 189–195. https://doi.org/10.1016/j.biortech.2012.06.041.

[38] Prescott, L.M., Harley, J.P., & Klein, D.A. (2002). " Microbiology." 5th ed. New York: McGraw Hill, 10-14. https://www.scirp.org/%28S%28351jmbntvnsjt1aadkposzje%29%29/reference/referencespapers.aspx?referenceid=2090198.

[39] Singh, R., Singh, P., & Sharma, R. (2014). "Microorganism as a tool of bioremediation technology for cleaning environment: A review." *International Academy of Ecology and Environmental Science,* 4, 1: 1-6. https://www.researchgate.net/publication/305386503_Microorganism_as_a_tool_of_bioremediation_technology_for_cleaning_environment_A_review.

[40] Misal, S.A., Lingojwar, D.P., Shinde, R.M., & Gawai, K.R. (2011). "Purification and characterization of azoreductase from alkaliphilic strains Bacillus badius." *Process Biochem.,* 46, 6: 264–269. https://doi.org/10.1016/j.procbio.2011.02.013.

[41] El, S., & Agathos, S.N. (2005) "Is bioaugmentation a feasible strategy for pollutant removal and site remediation?" *Current Opinion in Microbiology,* 8: 268-275. doi: 10.1016/j.mib.2005.04.011.

[42] Lee, C.M., Lu, C.J., & Chuang, M.S. (1994). "Effects of immobilized cells on the biodegradation of chlorinated phenols." *Water Sci Technol,* 30: 87–90. https://doi.org/10.2166/wst.1994.0449.

[43] Amin, A., Naik, A.T.R., Azhar, M., &Nayak, H. (2013). "Bioremediation of different wastewaters—a review." *Cont J Fish Aquat Sci.,* 7, 2: 7–17. https://www.researchgate.net/publication/318361794_BIOREMEDIATION_OF_DIFFERENT_WASTE_WATERS_-_A_REVIEW.

[44] Verma, J.P. &Jaiswal, D.K. (2016) "Book review: Advances in biodegradation and bioremediation of industrial waste." *Frontiers in Microbiology,* 6: 1-2. https://doi.org/10.3389/fmicb.2015.01555.

[45] Rao, A.G., Naidu, G.V., Prasad, K.K., Rao, N.C., Mohan, S.V., Jetty, A., & Sharma, P.N. (2004). "Anaerobic treatment of wastewater with high suspended solids from a bulk drug industry using fixed film reactors (AFFR)." *Bioresour Technol,* 93: 241–247. https://doi.org/10.1016/j.biortech.2003.11.004.

[46] Nandy, T., Shastry, S., & Kaul, S.N. (2002). "Wastewater management in a cane molasses distillery involving bio resource recovery." *J Environ Manage.,* 65: 25–38. https://doi.org/10.1006/jema.2001.0505.

[47] Khan, M.Z.H., & Mostafa, M.G. (2011) "Aerobic treatment of pharmaceutical wastewater in a biological reactor." *Int J Environ Sci.,* 1,7: 1797–1805. https://www.researchgate.net/publication/274721365_Aerobic_treatment_of_pharmaceutical_wastewater_in_a_biological_reactor.

[48] Altaf, M.S., & Ali, T.A. (2010). "Wastewater treatment using sequential batch reactor and development of microbiological method for the analysis of relative toxicity." *Pak J Nutr.,* 9, 6: 574–576. doi: 10.3923/pjn.2010.574.576.

[49] Wu, J., Muruganandham, M., Chang, L., Lee, G., Batalova, V., &Mokrousov, G. (2011). "Catalytic ozonation of oxalic acid using $SrTiO_3$ catalyst". Ozone: Science & Engineering, 33,1: 74-79. doi: 10.1080/01919512.2011.536903.

[50] Ghosh, M., Verma, S.C., Mengoni, A., &Tripathi, A.K. (2004). "Enrichment and identification of bacteria capable of reducing chemical oxygen demand of anaerobically treated spent wash." *J Appl Microbiol,* 6: 241–278. https://doi.org/10.1111/j.1365-2672.2004.02289.

[51] Chaturvedi, S., Chandra, R., & Rai, V. (2006). "Isolation and characterization of Phragmitisaustralis (L) rhizosphere bacteria from contaminated site for bioremediation of colored distillery effluent." *Ecol Eng.,* 27: 202–207. https://doi.org/10.1016/j.ecoleng.2006.02.008.

[52] Diaz, E. (2004). "Bacterial degradation of aromatic pollutants: a paradigm of metabolic versatility." *International microbiology: the official journal of the Spanish Society for Microbiology,* 7, 3: 173-180. PMID: 15492931. https://pubmed.ncbi.nlm.nih.gov/15492931/.

[53] Tewari, L., & Malviya, P. (2002). "Biodegradation of catechol by fluorescent Pseudomonas for sustainable environment". *J SciInd Res.,* 61: 70–74. http://nopr.niscair.res.in/bitstream/123456789/17653/1/JSIR%2061(1)%2070-74.pdf.

[54] Duffner, F.M., Kirchner, U., Bauer, M.P., & Muller, R. (2000). "Phenol/cresol degradation by the thermophilic Bacillus thermoglucosidasius A7: cloning and sequence analysis of five genes involved in the pathway." *Gene.,* 256: 215–221. https://doi.org/10.1016/S0378-1119(00)00352-8.

[55] Agarry, S.E., & Solomon, B.O. (2008). "Kinetics of batch microbial degradation of phenols by indigenous Pseudomonas fluorescence." *Int J Environ Sci Technol,* 5, 2: 223–232. https://doi.org/10.1007/BF03326016.

[56] Pino, N., &Peñuela, G. (2011). "Simultaneous degradation of the pesticides methyl parathion and chlorpyrifos by an isolated bacterial consortium from a contaminated site." *International Biodeterioration & Biodegradation,* 65, 6: 827-83. doi: 10.1016/j.ibiod.2011.06.001.

[57] Das, M.P., Bashwant, M., Kumar, K., & Das, J. (2012) "Control of pharmaceutical effluent parameters through bioremediation." *J Chem Pharm Res.,* 4, 2: 1061–1065. https://www.jocpr.com/articles/control-of-pharmaceutical-effluent-parameters-through-bioremediation.pdf.

[58] Bhattacharya, S.S., Syed, K., Shann, J., & Yadav, J.S. (2013). "A novel P450-initiated biphasic process for sustainable biodegradation of benzo pyrene in soil under nutrient-sufficient conditions by the white rot fungus Phanerochaetechrysosporium". *Journal of hazardous materials,* 261: 675- 683. https://doi.org/10.1016/j.jhazmat.2013.07.055.

[59] Jain, P.K., & Bajpai, V. (2012). "Biotechnology of bioremediation a review". *International Journal of Environmental Sciences,* 3,1: 535-549. https://www.researchgate.net/publication/311365994_Biotechnology_of_bioremediation-A_review.

[60] Chekroun, K.B., Sánchez, E., & Baghour, M. (2014). "The role of algae in bioremediation of organic pollutants." 1, 2: 19-32. https://www.researchgate.net/

publication/285887445_The_role_of_algae_in_bioremediation_of_organic_polluta nts.

[61] Tang, X., He, L.Y., Tao, X.Q., Dang, Z., Guo, C.L., Lu, G.N., & Yi, X.Y. (2010). "Construction of an artificial microalgal-bacterial consortium that efficiently degrades crude oil." *Journal of Hazardous Materials,* 181, 1–3, 1158-1162. doi: 10.1016/j.jhazmat.2010.05.033.

[62] de Lorenzo, V. (2009). "Recombinant bacteria for environmental release: what went wrong and what we have learnt from it." *Clin Microbiol Infect.,* 15: 63–65. https://doi.org/10.1111/j.1469-0691.2008.02683.x.

[63] Li, Q.S., Ogawa, J., Schmid, R.D., & Shimizu, S. (2001). "Engineering cytochrome P450 BM-3 for oxidation of polycyclic aromatic hydrocarbons." *Applied and environmental microbiology,* 67, 12: 5735-5739. doi: 10.1128/AEM.67.12.5735-5739.2001.

[64] Santoro, C., Arbizzani, C., Erable, B., & Ieropoulos, I. (2017a). "Microbial Fuel Cells: From Fundamentals to Applications. A Review." *J. Power Source.,* 356: 225–244. https://doi.org/10.1016/j.jpowsour.2017.03.109.

[65] Mohamed, A., Zmuda, H. M., Ha, P. T., Coats, E. R., & Beyenal, H. (2021). "Large-scale Switchable Potentiostatically Controlled/microbial Fuel Cell Bioelectrochemical Wastewater Treatment System." *Bioelectrochemistry,* 138, 107724: doi: 10.1016/j.bioelechem.2020.107724.

[66] Capodaglio, A. G., Molognoni, D., Puig, S., Balaguer, M. D., & Colprim, J. (2015). "Role of Operating Conditions on Energetic Pathways in a Microbial Fuel Cell." *Energ. Proceed.,* 74: 728–735. https://doi.org/10.1016/j.egypro.2015.07.808.

[67] Saratale, G. D., Saratale, R. G., Shahid, M. K., Zhen, G., Kumar, G., Shin, H.-S., Choi, Y. G. & Kim, S. H. (2017). "A Comprehensive Overview on Electro-Active Biofilms, Role of Exo-Electrogens and Their Microbial Niches in Microbial Fuel Cells (MFCs)." *Chemosphere,* 178, 534–547. doi: 10.1016/j.chemosphere.2017.03.066.

[68] Molognoni, D., Chiarolla, S., Cecconet, D., Callegari, A., & Capodaglio, A. G. (2018). "Industrial Wastewater Treatment with a Bioelectrochemical Process: Assessment of Depuration Efficiency and Energy Production." *Water Sci. Tech.,* 77, 1: 134–144. https://doi.org/10.2166/wst.2017.532.

[69] Bolognesi, S., Cecconet, D., Callegari, A., & Capodaglio, A. G. (2021a). "Bioelectrochemical Treatment of Municipal Solid Waste Landfill Mature Leachate and Dairy Wastewater as Co-substrates." *Environ. Sci. Pollut. Res.,* 28: 24639–24649. https://doi.org/10.1007/s11356-020-10167-7.

[70] Rafique, M., Sadaf, I. Rafique, M. S., & Tahir, M. B. (2017). "A review on green synthesis of silver nanoparticles and their applications." *Artificial Cells, Nanomedicine, and Biotechnology*, 45, 7: 1272–1291. https://doi.org/10.1080/21691401.2016.1241792.

[71] Anjum, M., Miandad, R., Waqas, M., Gehany, F., & Barakat, M. (2016). "Remediation of wastewater using various nano-materials." *Arabian Journal of Chemistry,* 12, 8: 4897-4919. https://doi.org/10.1016/j.arabjc.2016.10.004.

[72] Gutierrez, T. (2011) "Identifying polycyclic aromatic hydrocarbon-degrading bacteria in oil contaminated surface waters at Deepwater Horizon by cultivation,

stable isotope probing and pyrosequencing." *Rev Environ Sci Biotechnol,* 10, 4: 301-305. https://doi.org/10.1007/s11157-011-9252-9.

[73] Fang, H., Cai, L., Yang, Y., Ju, F., Li, X., Yu, Y., & Zhang, T. (2014). "Metagenomic analysis reveals potential biodegradation pathways of persistent pesticides in freshwater and marine sediments." *The Science of the total environment,* 470-471, 983-992. doi: 10.1016/j.scitotenv.2013.10.076.

Chapter 12

Metal Oxide Nanomaterials and Their Diverse Applications in Pharmaceuticals and Other Domains of Nanotechnology

Neetu Singh[1,*], Prabhat Kumar Singh[2], Ajay Kumar Saini[3] and Shalini Singh[4]

[1]Department of Physics, Govt. Degree College, Kuchlai-Sitapur, Uttar Pradesh, India
[2]Department of Physics, University of Lucknow, Uttar Pradesh, India
[3]Department of Chemistry, Govt. Degree College, Kuchlai-Sitapur, Uttar Pradesh, India
[4]Department of Zoology, Govt. Degree College, Kuchlai-Sitapur, Uttar Pradesh, India

Abstract

This chapter gives a broad review on different types of metal oxide nanomaterials, and their applications in various fields of nanotechnology. Basically, this chapter extensively gives the review on the current developments of metal oxide nanomaterials in the field of gas sensors using various gases like CO_2, O_3, NH_3, O_2, H_2S, CO, CH_4, H_2, NO_2, and LPG etc. The Gas sensing properties of different types of metal oxides nanomaterials towards particular target gases have been individually discussed in this chapter. This review also describes the major techniques and synthesis methods used in the preparation of metal oxide nanomaterials. Furthermore, the gas sensing mechanisms have also been discussed in detail. Apart from this, the different applications of metal oxide nanomaterials other than gas sensing have been discussed like in the field of pharmaceuticals, batteries, magnetic storage media, catalyst, various types of solar cells, energy conversion devices, rectifiers, electronic and optoelectronics devices.

* Corresponding Author's Email: neetusingh201989@gmail.com.

In: Pharmaceuticals: Boon or Bane
Editor: Neha Agarwal
ISBN: 979-8-88697-487-4

Keywords: metal oxide nanomaterials, gas sensors, pharmaceuticals, solar cell, batteries

1. Introduction

Metal oxide-based nanomaterials have grabbed the great attention of scientists and researchers due to their diversified area of applications. Metal oxides are the functional materials having a broad range of applications in chemistry, physics and material science [1- 4]. Metal oxides can be synthesized in very easy and are generally utilized the gas sensors [5, 6], batteries [7], biomedical [8], energy storage and conversion [9], microelectronics [10], catalytic purposes [11], solar cells [12], superconductors etc [13]. The size and surface morphology of metal oxides nanomaterials depend on the synthesis method, the pH value, heating temperature and time [14] and thus can be controlled during the growth process. The engineered metal oxides nanomaterials are the widest utilized nanomaterials because of their distinctive properties. The properties which make the metal oxide nanomaterials essential tools in the modern era of nanotechnology are their different optical properties, unique catalytic behavior, super paramagnetic properties, high ductility at raised temperatures, high surface area, high sensitivity and selective activity. For instance, the melting point of the nanomaterial is lower than that of a bulk material. Basically, nanotechnology is the study of the synthesis, design, characterizations and application of the different materials by controlling their shape and size at nanoscale level [15]. The nanoscale level ranges between 1-100 nm and materials pertaining in this range are called as nanomaterials. Several synthesis approaches are utilized in nanotechnology for the fabrication of nanomaterials. In general, there are two types of approaches for the synthesis of nanomaterials i.e., bottom- up and top- down approach. The top-down approach includes the grinding and crushing of a bulk material that providesvery fine particles. These fine particles are continuously crushed until they attain the nano-sized dimensions. On the other hand, the bottom-up approach assembles the atoms and molecules which lead to the formation of particles in the nano range. Figure 1 summarizes some major synthesis techniques used in nanotechnology for the preparation of nanomaterials. The bottom-up approach for synthesis of nano-sized materials involves the physical vapor deposition (PVD), pulsed laser deposition (PLD), chemical vapor deposition (CVD), atomic layer deposition (ALD), molecular beam epitaxy (MBE), spray pyrolysis method, sol-gel method and co-precipitation method etc. Whereas, top-down approaches include ball-milling technique, chemical etching and laser ablation synthesis techniques.

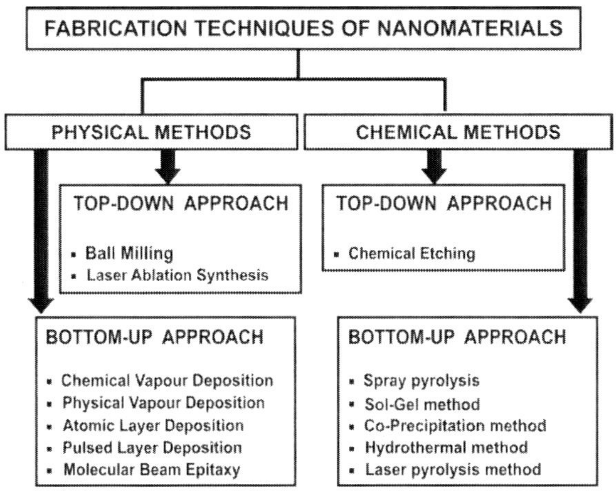

Figure 1. Techniques for the preparation of metal oxide nanomaterials [16].

Therefore, this chapter presents various types of application of metal oxide nanomaterials like pharmaceuticals, gas-sensors, batteries, solar cell, optoelectronic/electronic applications, etc. Many types of different metal oxide nanomaterials have been also discussed along with different methods to examine the changes in the different properties of the materials using different composition. It is not an easy task to cover the complete literature but some important parameters and the current literature related to nanomaterials have been discussed in this chapter. This chapter gives the basic overview for the researcher and scientists about the various classes of nanomaterials and their applications at one place.

2. Properties of Metal Oxide Nanomaterials

Currently, research in the field of metal oxide nanomaterials has emerged due to their ample potential applications in various areas [17]. The distinct physical and chemical properties of metal oxide nanomaterials are credited to their high surface area. The potential applications of metal oxide nanomaterials in the fields of medicine, agriculture, materials chemistry, energy, sensors, optical, electronics, catalysis, environment, information technology, biomedical etc. play a main role to grab the attention of researchers. Some changes in the properties have been observed by altering the size of some of the metal oxide nanomaterials like SnO_2, Al_2O_3, MgO, ZrO_2, AgO, CuO, ZnO, TiO_2, CeO_2 etc. As the size reaches to nano range, the surface area of the material is increased and interface atoms generate strain or stress which creates perturbations in the structure of the material [18]. The specific size of the nanomaterial can change the magnetic, structural, optical, morphological, conducting, chemical, and electronic properties [19]. Iron oxides have attracted a lot of attention from researchers as they are associated with numerous potential applications such as magnetic storage devices (MSD), magnetic resonance imaging (MRI) and contrast agents [20]. Size-dependency was observed in γ-Fe_2O_3 nanomaterials, where the nanomaterials possessing the size 55nm show the ferromagnetic behavior and the material having size 12 nm show superparamagnetic behavior [21]. Advanced methods of preparing required sized metal oxide nanomaterials are essential for achieving all the desired magnetic, physical, structural, optical, electronic, and chemical properties. Electrical properties (conductance and resistance) strongly depend on the size of these nanomaterials, especially for the metal oxides like SnO_2, WO_3, ZnO, CuO, MgO, and In_2O_3 [22] in gas sensing applications. Electrical conductivity is an important property of TiO_2 nanomaterials that makes it useful in the areas of sensors, optoelectronics, energy storage devices and photovoltaics. The size of TiO_2 nanomaterial can be easily reduced at high temperatures which directly affects its conductivity [23]. ZnO nanomaterials with the wurtzite-type structure have large number of potential applications from gas sensors to varistors [24]. CuO is used as a redox catalyst in various oxidation processes. Apart from this, CuO is also used in the several applications such as microwave irradiation, photoconductors and photothermal [25]. MgO nanomaterials are hugely applied in different types of chemical industries for reducing the gaseous air pollutants (CO, NOx, CO_2, SOx) and it is also used as a catalyst in organic synthesis [26]. Al_2O_3 nanomaterial is used as a supportive material for active phases in the field of catalysis [27]. ZrO_2 nanomaterials are used as a strong electrolyte, catalysis and potential gas sensing materials as well due to its dynamic properties [28]. CeO_2 nanoparticles are used in various fields like electrochemistry, gas sensing, biomedical and catalysis etc [29]. TiO_2 nanoparticles are used as photocatalysts for pollutant elimination and solar cells etc [30]. ZnO has a wide band gap which is largely used in several fields as UV-blocker in sunscreens, solar cells, mixed varistors, gas sensors, laser sources, and as catalysts for several kinds of organic and inorganic reactions [24]. All the aforementioned properties and applications of the metal oxide nanomaterials closely depend on size, shape and structure which we obtained. Thus, it is very much clear that numerous factors will play the role to affect the properties and the applications of metal oxide nanomaterials. Different types of applications and properties of some metal oxide nanomaterials have been shown in the tabular form in Table 1 [31-42].

Table 1. Some significant applications and properties of metal oxide nanomaterials

S. No.	Type of metaloxide	Application	Properties	Ref.
1.	TiO_2	Biomedical and Environmental	Electronic	31-33
2.	Co_3O_4	Energy storage	Optical and magnetic	34
3.	SnO_2	Biomedical	Electronic and optical	35
4.	Fe_2O_3,Fe_3O_4	Biomedical	Magnetic	36-37
5.	Bi_2O_3	Water treatment	Optoelectronic	38
6.	CuO,Cu_2O	Antimicrobial agent	Optical and electronic	39
7.	SiO_2	Biomedical and cosmetics	Biocompatible	40
8.	ZrO_2	Catalysis	Electro-optical, dielectric	41
9.	CeO_2	Antioxidant effects, biomedical	Electronic	42

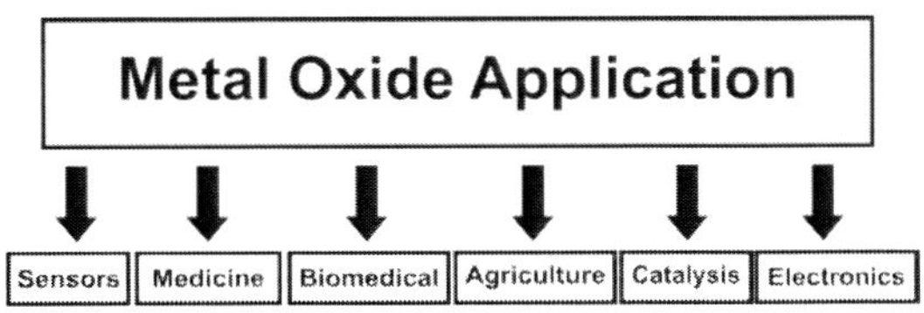

Figure 2. Various applications of metal oxides.

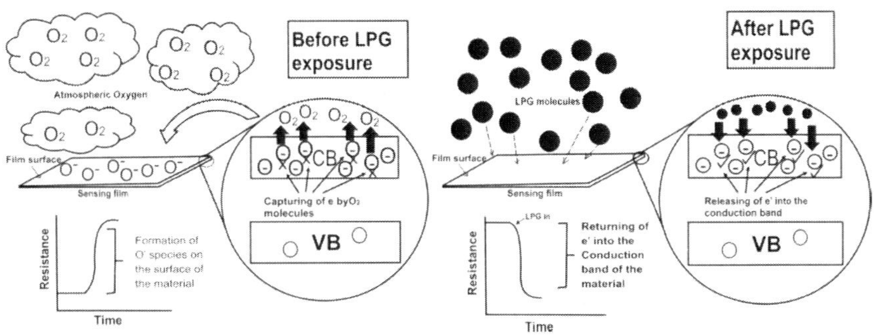

Figure 3. Pictorial diagram for liquefied petroleum gas sensing mechanism for metal oxides [16].

3. Diverse Applications of Metal Oxide Nanomaterials

Metal oxide nanomaterials are being utilized in several applications due to their dynamic behavior. Figure 2 shows different applications of metal oxides and some of them have been discussed in detail here.

3.1. Gas Sensors

Over the past decades, metal oxide nanomaterials are the highly exploited sensing materials for the fabrication of gassensors due to its fast response, low cost, easy handling, a wide variety of target gases, high stability and longer life time etc. However, they have some disadvantages also like high erenergy consumption and high operation temperature. Having large surface area, greater active sites, high porosity and a very high surface reactivity, metal oxide nanomaterials are being applied i n the domain of gas sensing [4, 43]. Basically, a gas sensor that utilizes the metal oxide nanomaterial consists of a heating layer for maintaining the working temperature, conducting electrodes for measuring the resistance and sensing film of the material which varies its resistance after the exposure of the target gas as shown in Figure 3.

Gas sensing mechanism is explained by the number of active surface sites and adsorption of oxygen species at the surface of the material that further enhances the number of active sites [44]. Gas sensors are necessary as they could prevent us from a disastrous situation by sensing the dangerous gas on the leakage of that gas. A good sensor depends on various factors such as high sensitivity about the target gas, fast response time, lesser energy consumption, high stability and fast recovery time [45]. The metal oxide nanomaterials are being utilized for the fabrication of different type of gas sensors like LPG (Liquified Petroleum Gas) sensor, Carbon Dioxide (CO_2) sensor, Nitrogen Dioxide (NO_2) sensor, Ammonia (NH_3) sensor, H_2S sensor, Oxygen sensor, Ozone gas sensor and CO (Carbon Monoxide) sensor etc.

3.2. Batteries

Rechargeable batteries grabbed a great amount of attention from the researchers in the past few years due to some environmental factors. Lithium-ion batteries have played a vital role in our day-to-day life and these are being vastly utilized in the portable electronic devices, hybrid vehicles and electrically charged vehicles because of their very long life. Scientists are putting their efforts to enhance the performance of Lithium-ion batteries by using metal oxide nanomaterials. Many researchers have found that metal oxide nanomaterials have increased the capability of Lithium-ion batteries in terms of durability. However, some nano-metal oxides have very low electrical conductivity but that could be increased by adding the electrolytes in metal oxides [46], or adding a thin layer of conducting polymer in it [47].

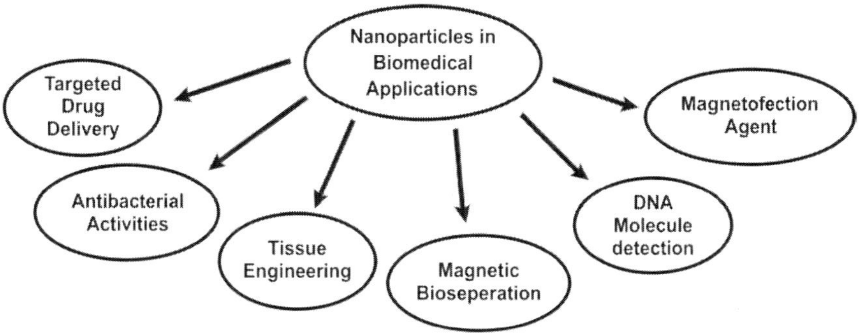

Figure 4. Various biomedical applications of metal oxide nanomaterials.

3.3. Solar Cells

The prime function of a solar cell is to generate the charge carriers in a material that absorbs the photons and to separate these carriers from the conductive materials that transmit the electricity. Charge carriers are generated mostly near the surface of the material. When a photon is incident upon a semiconductor or a material, the pairs of electron-holes are generated which increase the conductivity of the material. Metal oxide semiconductors are eco-friendly, cost-effective and most importantly, highly stable materials. Over the several decades, they are being used in photovoltaics, in dye solar cells as photoelectrodes and also used to design the metal oxide p–n junctions [48]. The remarkable flexible properties like environment friendly, cost-effectiveness and easy synthesis make metals oxides a demanding material in the domain of solar cells. Having large band gap and large band gap energy range, metal oxides are also used in photo-harvesters. Semiconducting materials are the materials which can absorb the solar energy and these materials use this energy for the transformation of electrons from valence band to conduction band. This passage of electrons from valence band to conduction band creates an electron hole pair in the materials, and consequently this process is utilized to convert the electric energy from the solar energy. Various metal oxides such as ZnO, TiO_2, CuO etc. have good photovoltaic properties with a large band gap. These metal oxides can easily be prepared in the form of nanomaterials and these nanomaterials can further be utilized to design the solar cells [49]. Amongst all the metal oxide nanomaterials, the copperoxides (Cu–O) are the most stable and useful material till date in the field of photovoltaics. The band gap energy of copperoxides ranges between 1.4 and 2.2 eV [50]. Cuprous oxide (Cu_2O) is used as photovoltaic material because of its cost-effectiveness, good mobility, high stability, easy synthesis, eco-friendliness, great availability, and no toxicity.

3.4. Biomedical

Various metal oxides such as $NiFe_2O_4$, $CoFe_2O_4$ and Fe_3O_4 are the potential nanomaterials in different biomedical applications from magnetic hyperthermia applications to targeted drug delivery [51]. It is very essential for this application that the nanomaterials that are being used must be biocompatible. Along with biocompatibility, water dispersibility is another most important parameter for these nanomaterials. By reducing the size of magnetic nanomaterials, they can be changed into superparamagnetic nanomaterials and can be utilized in medical applications like magnetic hyperthermia. If, the surface of the nanomaterials is coated by different type of ligands then this water dispersibility of nanomaterial can be attained [52,53]. Various possible biomedical applications of nanomaterials have been shown in Figure 4.

Conclusion

Some important applications of metal oxide nanomaterials such as gas sensors, batteries, solar cells and biomedical engineering have been explained through this chapter. Metal oxide nanomaterials with their various nano structures are proved to be important materials to fabricate gas sensors with high sensitivity, high stability, good sensing response and recovery time. Some improvement can be done in commercially available gas sensors by controlling the temperature, size, structure and composition of the nanomaterials. Metal oxide nanomaterials have also proved to be excellent materials in photovoltaic applications such as solar cells. The solar cells designed by using metal oxides are more stable and consume less energy. In the domain of batteries, metal oxides have been given fruitful results as the batteries made by using these materials have a good lifetime in comparison to other materials batteries. Metal oxide nanomaterials have shown its capability in the pharmaceutical industries as well. This chapter could be beneficial for the researchers working on metal oxide nanomaterials.

References

[1] Henrich, V.E. (1985). "The surface of metaloxides." *Rep.Prog.Phys.,*48,11:1481. https://doi.org/10.1088/0034-4885/48/11/001.

[2] Meyer, J., Hamwi, S., Kroger, M., Kowalsky, W., Riedl, T., & Kahn, A. (2012). "Transition metaloxides for organic electronics: energetics, device physics and applications." *Adv. Mater.*, 24, 40:5408-27. doi: https://doi.org/10.1002/adma.201201630.

[3] Pinna, N., Hochepied, J.F., Niederberger, M., & Gregg, M. (2009). "Chemistry and physics of metal oxide nanostructures." *Phys. Chem. Chem. Phys.*, 11, 19:3607. doi: https://doi.org/10.1039/b905768d.

[4] Shen, G., Chen, P.C., Ryua, K., & Zhou, C. (2009). "Devices and chemical sensing applications of metal oxide nanowires." *J. Mater. Chem.*, 19, 828-839. https://doi.org/10.1039/B816543B.

[5] Lupan, O., Braniste, T., Deng, M., Ghimpu, L., Paulowicz, I., Mishra, Y.K., Kienle, L., Adelung, R., & Tiginyanu, I. (2015). "Enhanced ethanol vapor sensing performances of copper oxide nanocrystals with mixed phases." *Sensors and Actuator B-Chem.*, 224:434-448. https://doi.org/10.1016/j.snb.2015.10.042.

[6] Singh, N., Singh, P.K., Singh, M., Tandon, P., Singh, S.K., & Singh,S. (2019). "Fabrication and characterization of polyaniline, polyaniline/MgO (30%) and polyaniline/MgO (40%) nanocomposites for their employment in LPG sensing at room temperature." *J Mater Sci: Mater Electron,* 30: 4487–4498. https://doi.org/10.1007/s10854-019-00737-y.

[7] Cao, K., Jiao, L., Liu, Y., Liu, H., Wang, Y., & Yuan, H. (2015). "Ultra High Capacity Lithium-ion Batteries with Hierarchical CoO Nanowire Clusters as Binder Free Electrodesa." *Adv Funct Mater.,* 25:1082–1089. https://doi.org/10.1002/adfm.201403111.

[8] Holzinger, M., LeGoff, A., & Cosnier, S. (2014). "Nanomaterials for Biosensing Applications: A Review." *Front. Chem.*, 2,1. https://doi.org/10.3389/fchem.2014.00063.

[9] Zhang, Q., Uchaker, E., & Candelaria, S.L. (2013). "Nanomaterials for Energy Conservation and Storage." *Chem. Soc. Rev.*, 42:3127-3171. https://doi.org/10.1039/C3CS00009E.

[10] Nourozi, B., Aminian, A., Fili, N., Zangeneh, Y., Boochani, A., & Darabi, P. (2019). "The Electronic and Optical Properties of MgO-nanolayer: Based on GGA, Mbj." *Result in Phys.*, 12, 2038-2043. https://doi.org/10.1016/j.rinp.2019.02.054.

[11] Wellia, D.V., Kusumawati, Y., Diguna, L.J., & Amal, M.I. (2017). "Introduction of nanomaterials for photocatalysis." In *Nanocomposites for visible light-induced photocatalysis,* edited by Khan, M., Pradhan, D., & Sohn, Y. Springer, Cham, pp.1–17. https://www.academia.edu/34887141/Nanocomposites_for_Visible_Light_induced_Photocatalysis_Book_.

[12] Grätzel, M. (2014). "The Light and Shade of Perovskite Solar Cells." *Nat. Mater.*, 13, 838-842. https://doi.org/10.1038/nmat4065.

[13] Sidorov, N.S., Palnichenko, A.V., & Vyaselev, O.M. (2012). Superconductivity in Mg/MgO Interface." *Phys. C, Superconductivity.,* 480:123-125. https://doi.org/10.1016/j.physc.2012.05.003.

[14] Laska, M., Valtyni, J., & Fellner, P. (1993). "Influence of pH on the Crystal Size Distribution of $Mg(OH)_2$ Prepared by the Hydration of MgO." *Cryst. Res. Technol.*, 28:931. https://doi.org/10.1002/crat.2170280709.

[15] Zhang, L., & Webster, T.J. (2009). "Nanotechnology and nanomaterials: Promises for improved tissue regeneration." *Nano Today,* 4:66-80. https://doi.org/10.1016/j.nantod.2008.10.0.

[16] Singh, P.K. (2021). "*Synthesis Characterizations and LPG sensing application of nanostructured metaloxides*." PhD Thesis, University of Lucknow, India.

[17] Taylor, R., Coulombe, R.S., Otanicar, T., Phelan, P., Gunawan, A., Lv, W., Rosengarten, R., Prasher, R., & Tyagi, H. (2013). "Small particles, big impacts: A review of the diverse applications of nanofluids." *AIP Journal of Appl Phys.,* 113, 1:011301. https://doi.org/10.1063/1.4754271.

[18] Bansal, V., Poddar, P., Ahmad, A., & Sastry, M. (2006). "Room-Temperature Biosynthesis of Ferroelectric Barium Titanate Nanoparticles." *J Am Chem Soc.,* 128, 36:11958-11963. https://pubs.acs.org/doi/10.1021/ja063011m.

[19] Liu, W.T. (2006). "Nanoparticles and their biological and environmental applications." *J Biosci Bioeng.,* 102, 1:1–7.doi: https://doi.org/10.1263/jbb.102.1.

[20] Lang, C., Schuler, D., & Faivre, D. (2007). "Synthesis of Magnetite Nanoparticles for Bio-and Nanotechnology: Genetic Engineering and Biomimetics of Bacterial Magnetosomes." *Macromol Biosci,* 7:144–151. https://doi.org/10.1002/mabi.200600235.

[21] Jun, Y.W., Seo, J.W., & Cheon, J. (2008). "Nanoscaling laws of magnetic nanoparticles and their applicabilities in biomedical sciences." *Acc Chem Res.,* 41, 2:170–189. doi: https://doi.org/10.1021/ar700121f.

[22] Klem, M.T., Resnick, D.A., Gilmore, K., Young, M., Idzerda, Y.U., & Douglas, T. (2007). "Synthetic Control over Magnetic Moment and Exchange Biasin All-Oxide Materials Encapsulated within a Spherical Protein Cage." *J Am Chem Soc.,* 129, 1:197–201. https://doi.org/10.1021/ja0667561.

[23] Kubaschewski, O. (1982). *"The oxide handbook, 2nd Edition."* Edited by G.V. Sansonov, IFI/Plenum Press, New York. https://doi.org/10.1002/bbpc.19820860820.

[24] Nuraje, N., Su, K., Haboosheh, A., Samson, J., Manning, E.P., Yang, N.I., & Matsui, H. (2006). "Room Temperature Synthesis of Ferroelectric Barium Titanate Nanoparticles Using Peptide Nanorings as Templates." *Adv Mater.,* 18, 6:807–811. doi: https://doi.org/10.1002/adma.200501340.

[25] Klem, M.T., Mosolf, J., Young, M., & Douglas, T. (2008). "Photochemical Mineralization of Europium, Titanium, and Iron Oxyhydroxide Nanoparticles in the Ferritin Protein Cage." *Inorg Chem.,* 47, 7:2237–2239. https://doi.org/10.1021/ic701740q.

[26] Zhang, W., Zhang, D., Fan, T., Ding, J., Guo, J.Q., & Ogawa, H. (2006). "Fabrication of ZnO microtubes with adjustable nanopores on the walls by the templating of butterfly wing scales." *Nanotechnology,* 17:840–844. https://doi.org/10.1088/0957-4484/17/3/038.

[27] Aizenberg, J., Hanson, J., Koetzle, T.F., Weiner, S., & Addadi, L. (1997). "Control of Macromolecule Distribution within Synthetic and Biogenic Single Calcite Crystals." *J Am Chem Soc.,* 119, 5:881–886. https://doi.org/10.1021/ja9628821.

[28] Biró, L.P., Kertész, K., Vértesy, Z., Márk, G.I., Bálint, Z., Lousse, V., & Vigneron, J.P. (2007). "Living photonic crystals: butterfly scales – nanostructure and optical properties." *Materials Science and Engineering: C,* 27(5-8):941-946. https://doi.org/10.1016/j.msec.2006.09.043.

[29] Zou, D., Xu, C., Luo, H., Wang, L., & Ying, T. (2008). "Synthesis of Co3O4 nanoparticles via an ionic liquid-assisted methodology at room temperature." *Mater Lett.,* 62(12-13):1976–1978. https://doi.org/10.1016/j.matlet.2007.10.056.

[30] Wilkes, J.S., & Zaworotko, M.J. (1992). "Air and water stable 1-ethyl-3-methylimidazolium based ionic liquids." *Chem Commun.,* 13:965–967. https://doi.org/10.1039/C39920000965.

[31] Jafari, S., Mahyad, B., Hashemzadeh, H., Janfaza, S., Gholikhani, T., & Tayebi, L. (2020). "Biomedical applications of TiO_2 nanostructures: recent advances." *Int J Nanomedicine*, 15:3447-3470. doi: https://doi.org/10.2147/IJN.S249441.

[32] Goh, P.S., Ng, B.C., Lau, W.J., & Ismail, A.F. (2015). "Inorganic Nanomaterials in Polymeric Ultra filtration Membranes for Water Treatment." *Sep. Purif. Rev.*, 44:216–249. https://doi.org/10.1080/15422119.2014.926274.

[33] Cao, Y., Yuan, F., Yao, M., Bang, J.H., Lee, J.H. (2014). "A new synthetic route to hollow Co3O4 octahedra for supercapacitor applications." *Crst Eng Comm.,* 16:826–833. https://doi.org/10.1039/C3CE41840E.

[34] Vidhu, V.K., & Philip, D. (2015). "Biogenicsynthesis of SnO_2 nanoparticles: evaluation of antibacterial and antioxidant activities." *Spectrochim. Acta A Mol. Biomol. Spectrosc.*, 134, 372–379. doi: https://doi.org/10.1016/j.saa.2014.06.131.

[35] Molina, M.M., Seabra, A.B., deOliveira, M.G., Itri, R., & Haddad, P.S. (2013). "Nitric oxide donor superparamagnetic iron oxide nanoparticles." *Mater. Sci. Eng. C.,* 33, 2:746-751. doi: https://doi.org/10.1016/j.msec.2012.10.027.

[36] Laurent, S., Forge, D., Port, M., Roch, A., Robic, C., Vander, Elst, L.V., Muller, R.N. (2008). "Magnetic iron oxide nanoparticles: synthesis, stabilization,

vectorization, physicochemical characterizations, and biological applications." *Chem. Rev.*, 108, 6:2064–2110. doi: https://doi.org/10.1021/cr068445e.

[37] Hou, J., Yang, C., Wang, Z., Zhou, W., Jiao, S., & Zhu, H. (2013). "Insitu synthesis of α–β phase hetero junction on Bi2O3 nanowires with exceptional visible-light photocatalytic performance." *Appl. Catal. B Environ.*,142-143,504–511. https://doi.org/10.1016/j.apcatb.2013.05.050.

[38] Rubilar, O., Rai, M., Tortella, G., Diez, M.C., Seabra, A.B., & Dura, N. (2013). "Biogenic nanoparticles: copper, copper oxides, copper sulphides, complex copper nanostructures and their applications." *Biotechnol. Lett.,* 35, 9:1365–1375.doi: https://doi.org/10.1007/s10529-013-1239-x.

[39] Zhu, Y.Y., & Liao, L.M. (2015). "Applications of Nanoparticles for Anticancer Drug Delivery: A Review." *J. Nanosci. Nanotechnol.*, 15, 7:4753–4773. https://doi.org/10.1166/jnn.2015.10298.

[40] Bansal, V., Rautaray, C., Ahmad, A., & Sastry, M. (2004). "Biosynthesis of zirconia nanoparticles using the fungus *Fusarium oxysporum*." *J. Mater. Chem.*, 14:3303–3305. https://doi.org/10.1039/B407904C.

[41] Tarnuzzer, R.W., Colon, J., Patil, S., & Seal, S. (2005). "Vacancy engineered cerianano structures for protection from radiation-induced cellular damage." *Nano Lett.,* 5, 12:2573–2577. doi: https://doi.org/10.1021/nl052024f.

[42] Liu, X., Zhang, J., Wu, S., Yang, D., Liu, P., Zhang, H., Wang, S., Yao, X., Zhu, G., & Zhao, H. (2012). "Single crystal α-Fe2O3 with exposed {104} facets for high performance gas sensor applications." *RSC Adv.,* 2:6178–6184. https://doi.org/10.1039/C2RA20797D.

[43] Li, X., Wei, W., Wang, X., Kuai, L., & Geng, B. (2011). "Single-crystalline α-Fe2O3 oblique nanoparallelepipeds: High-yield synthesis, growth mechanism and structure enhanced gas-sensing properties." *Nanoscale*, 3, 718–724. https://doi.org/10.1039/C0NR00617C.

[44] Ayesh, A.I., Abu-Hani, A.F.S., Mahmoud, S.T., & Haik, Y. (2016). "Selective H2S sensor based on CuO nanoparticles embedded inorganic membranes." *Sens Actuators B Chem.,* 231:593–600. https://doi.org/10.1016/j.snb.2016.03.078.

[45] Liu, X., Cheng, S., Liu, H., Sha, H., Zhang, H.D., Ning, H.A. (2012). "A Survey on Gas Sensing Technology." *Sensors*, 12:9635–9665. https://doi.org/10.3390/s120709635.

[46] Jeong, Y.U., & Manthiram, A. (2002). "Nanocrystalline Manganese Oxides for Electrochemical Capacitors with Neutral Electrolytes." *J Electrochem Soc.,* 149, 11:1419–1422.doi: https://doi.org/10.1149/1.1511188.

[47] Xiao, W., Chen, J.S., Lu, Q., & Lou, X.W. (2010). "Porous Spheres Assembled from Polythiophene(PTh)-Coated Ultrathin MnO2 Nanosheets with Enhanced Lithium Storage Capabilities." *J Phys Chem C.,* 114, 27:12048–12051. https://doi.org/10.1021/jp104227e.

[48] Jose, R., Thavasi, V., Ramakrishna, S. (2009). "Metal Oxides for Dye-Sensitized Solar Cells." *J Am Ceram Soc.,* 92, 289–301. https://doi.org/10.1111/j.1551-2916.2008.02870.x.

[49] Wibowo, A., Marsudi, M.A., Amal, M.I., Ananda, M.B., Stephanie, R., Ardy, H., Diguna, L.J. (2020). "ZnO nanostructured materials for emerging solar cell applications." *RSC Adv.*, 10:42838–42859. https://doi.org/10.1039/D0RA07689A.

[50] Anderson, A.Y., Bouhadana, Y., Barad, H.N., Kupfer, B., Rosh-Chodesh, E., Aviv, H., Tischler, Y.R., Rühle, S., & Zaban, A. (2014). "Quantum Efficiency and Bandgap Analysis for Combinatorial Photovoltaics: Sorting Activity of Cu–O Compounds in All-Oxide Device Libraries." *ACS Comb Sci.,* 16:53–65. https://doi.org/10.1021/co3001583.

[51] Munjal, S., Khare, N., Sivakumar, B., & Sakthikumar, D.N. (2019). "Citric acid coated CoFe2O4 nanoparticles transformed through rapid mechanochemical ligand exchange for efficient magnetic hyperthermia applications." *J. Magn, & Magn. Mater.,* 477:388–395. https://doi.org/10.1016/j.jmmm.2018.09.007.

[52] Munjal, S., & Khare, N. (2017). "Transforming single domain magnetic CoFe2O4 nanoparticles from hydrophobic to hydrophilic by novel mechanochemical ligand exchange." *J. Nanoparticle Res.*, 19,18. https://doi.org/10.1007/s11051-016-3700-y.

[53] Munjal, S., & Khare, N. (2016). "Cobalt ferrite nanoparticles with improved aqueous colloidal stability and electrophoretic mobility." *AIP Conference Proceedings*, 1724: 020092, doi: https://doi.org/10.1063/1.4945212.

Index

A

B

C

I

L

M

N

O

P

Editor's Contact Information

Dr. Neha Agarwal

Head of the Department of Chemistry

Navyug Kanya Mahavidyalaya

Rajendra Nagar, Lucknow, Uttar Pradesh, India

Email: nehaagarwal4074@gmail.com.